Calculus Explorations

Paul A. Foerster

KEY CURRICULUM PRESS
Innovators in Mathematics Education

Editors: Sarah Block and Bill Medigovich
Production Editors: Deborah Cogan and Jason Luz
Copyeditor: Joseph Siegel
Editorial Assistant: James A. Browne
Compositors: Esther Adams and E. A. Pauw
Mathematics Reviewer: Cavan Fang
Cover Design: Mary Ann Ohki
Design Coordinator: Diana Krevsky

Publisher: Steven Rasmussen
Editorial Director: John Bergez

Key Curriculum Press
1150 65th Street
Emeryville, CA 94608
510-595-7000
editorial@keypress.com
http://www.keypress.com

Printed in the United States of America

10 9 8 7 6 5 4 3 2   05 04 03 02

ISBN 1-55953-311-0

The graphs in this text were created using PSMathGraphs II. PSMathGraphs is a trademark of MaryAnn Software.

Contents

The Calculus of Exponential and Logarithmic Functions

The Calculus of Growth and Decay

The Calculus of Plane and Solid Figures

Algebraic Calculus Techniques for the Elementary Functions

The Calculus of Motion—Averages, Extremes, and Vectors

The Calculus of Variable-Factor Products

The Calculus of Functions Defined by Power Series

To the Instructor

The materials in *Calculus Explorations* are designed to give students hands-on practice with the concepts and techniques of calculus. Most of the materials are intended to allow students to discover a concept, on their own or in cooperative groups, before the concept is reinforced by classroom discussion or lecture.

The Explorations can be used with any calculus text. They are particularly suited to "reform" texts that make use of graphing calculators, which allow students to learn by graphical and numerical methods as well as by traditional algebraic methods. Writing forms a strong component of the Explorations because students are expected to verbalize their conjectures and conclusions and to write a paragraph describing what they learned as a result of doing an Exploration. The sequence of topics in the Explorations follows that of Paul Foerster's *Calculus: Concepts and Applications* (Key Curriculum Press, 1998), and *Calculus Explorations* includes an extensive index of major topics to guide you to Explorations suitable for your immediate need.

The title *Calculus Explorations* has been chosen deliberately. Students learn the concepts by exploring them on their own, without having read the materials in their calculus text. Avoid calling the Explorations "worksheets" or "reviews" because in most students' minds these names carry unpleasant or boring connotations. Some instructors call the Explorations **Games,** alluding to the fact that there is some sport involved and to the idea that they are not necessarily done for a grade. (It is not whether you win or you lose. It is how you play the *Game*!) Most of the Explorations consist of a single sheet. Space is provided for students to work right on the sheet. This feature is particularly important when something must be drawn on an accurate graph.

The Explorations can be used both in class and outside of class. For classroom use you can ask your students to work in **cooperative groups** for fifteen to thirty minutes, with minimal guidance from you, then follow up with five or ten minutes of instruction to make sure they have not "discovered" something incorrect. You can also use the Explorations as **directed discovery exercises** in which students work the problems in the Explorations one at a time, on their own or in groups, with follow-up answers and discussion after each problem. After following up on a particular problem, you may say to students, "Okay, you have three minutes to do Problem 5." Or, the materials in the Explorations can be used as a basis for your **lecture** on a particular topic.

For out-of-class work you can assign an Exploration as homework, then follow up on it during the next class period. Students can do the Exploration on their own or can meet with other students outside of class and work cooperatively.

For out-of-class or for in-class work, it is helpful to ask students to research in their textbooks and get more information about the particular topic of an Exploration. Some students will think they are outsmarting you by reading the text *first,* then working the Exploration. Actually, you are outsmarting them by getting *them* to read the book!

Taken as a whole, the Explorations uncover virtually the entire calculus course for your students. You could almost consider your calculus text a supplement to the Explorations. However, be careful not to go overboard with this means of instruction. A variety of instruction methods allows you to address students' different learning styles. Sometimes the traditional lecture is the appropriate choice for a topic, both to be time efficient and to prepare students for future courses in which the lecture method may be used more extensively.

Procedurally, students can purchase or be issued copies of *Calculus Explorations*, or you can reproduce the Explorations one at a time as needed and distribute them to your students. The former saves you the time of photocopying and allows students to look up answers at the back of the book, which is an advantage if they are working on their own. The latter gets students away from the "workbook stigma," allowing them to concentrate on one Exploration at a time. The fact that answers are *not* available when using this procedure may better accommodate your objectives for the Explorations.

The last question in an Exploration is often "What did you learn as a result of doing this Exploration that you did not know before?" You will find it interesting to read student responses to this question. Sometimes you will learn surprising things about your students or about their insights into mathematics. Often they have something mixed up. Sometimes the comments include personal information that is helpful for you to know. Occasionally you will learn something about calculus that you did not realize yourself!

To the Student

This book contains exercises called Explorations. Each one concerns a particular concept or technique in calculus. These Explorations are designed to be done with minimal guidance from your instructor. Although you may do them on your own, they are most effective when you work on them with others in cooperative groups. By working in a group, you can share your ideas and get feedback about whether or not a conjecture you have made is correct.

There are at least four methods you can use to learn calculus. Traditionally, there was only the **algebraic** method. With this method, calculus involved formulas for derivatives, integrals, volumes, and so on. Now you can also use **numerical** and **graphical** methods. For instance, the concept of limit comes across very clearly as you explore a table of numbers that get closer and closer to some fixed value. Graphs show discontinuities and the location of maximum or minimum values of a function. Accurately drawn graphs enable you to see concepts such as local linearity or the fact that a line with slope equal to the derivative is really tangent to the graph. Numerical and graphical methods are usually faster than algebraic methods, so when you have worked through problems in an Exploration, you will have time left over to write about what you have learned as a result of doing the Exploration. The Explorations ask you to use **verbal** methods to present your conclusions and to describe what you have learned.

If you are reading this introduction, you most likely have the entire book of Explorations on hand. Solutions are included at the end of this book. The most effective way to use the solutions is to do an entire Exploration first, then make sure your answers agree. Do not consider the Exploration finished until you understand why the given solution is correct and why your original thinking may have been in error.

You can find additional information about the topic of a given Exploration by looking at the index in this book. To confirm that what you discovered in an Exploration is correct, use the index to find more information about the topic of the Exploration. Or, first read about the topic in your calculus text, then try working through the Exploration. Try various approaches and see which works best for you.

Best wishes for your study of calculus!

Name _____ Group _____ Date _____

Exploration 1: Instantaneous Rate of Change of a Function

<u>Objective</u>: Explore the instantaneous rate of change of a function.

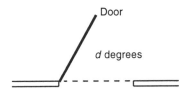

The diagram shows a door with an automatic closer. At time $t = 0$ seconds someone pushes the door. It swings open, slows down, stops, starts closing, then slams shut at time $t = 7$ seconds. As the door is in motion the number of degrees, d, it is from its closed position depends on t.

1. Sketch a reasonable graph of d versus t.

2. Suppose that d is given by the equation

   $d = 200t \cdot 2^{-t}$.

   Plot this graph on your grapher. Sketch the results here.

3. Make a table of values of d for each second from $t = 0$ through $t = 10$. Round to the nearest $0.1°$.

t	d
0	
1	
2	
3	
4	
5	
6	
7	
8	
9	
10	

4. At time $t = 1$ second, does the door appear to be opening or closing? How do you tell?

5. What is the average rate at which the door is moving for the time interval [1, 1.1]? Based on your answer, does the door seem to be opening or closing at time $t = 1$? Explain.

6. Find an estimate of the *instantaneous* rate at which the door is moving at time $t = 1$ second. Show how you get your answer.

7. In calculus you will learn by four methods:
   • algebraically,
   • numerically,
   • graphically,
   • verbally (talking and writing).
   What did you learn as a result of doing this Exploration that you did not know before?   (Over)

Name _____ Group _____ Date _____

Exploration 2: Graphs of Functions

Objective: Recall the graphs of familiar functions, and tell how fast the function is changing at a particular value of x.

For each function:
  a. Plot the graph on the axes provided. You may use tables or grapher plots first if necessary.
  b. Tell whether the function is increasing, decreasing, or not changing when $x = 1$. If it is increasing or decreasing, tell whether the rate of change is slow or fast.

1. $f(x) = 3^{-x}$

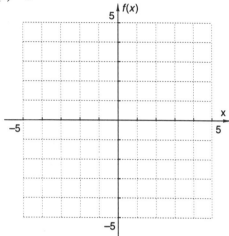

2. $f(x) = \sin \frac{\pi}{2} x$

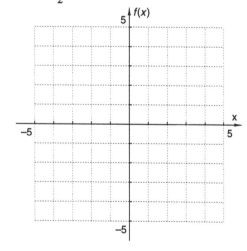

3. $f(x) = x^2 + 2x - 2$

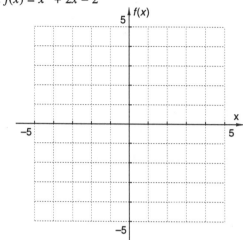

4. $f(x) = \sec x$

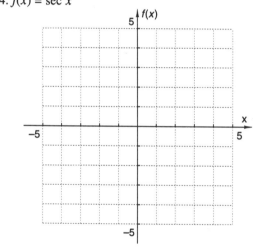

5. $f(x) = \frac{1}{x}$

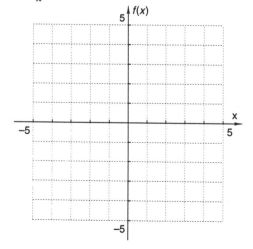

Calculus Explorations
© 1998 Key Curriculum Press

Exploration 3: Introduction to Definite Integrals

Objective: Find out what a definite integral is by working a real-world problem that involves the speed of a car.

As you drive on the highway you accelerate to 100 feet per second to pass a truck. After you have passed, you slow down to a more moderate 60 ft/sec. The diagram shows the graph of your velocity, $v(t)$, as a function of the number of seconds, t, since you started slowing.

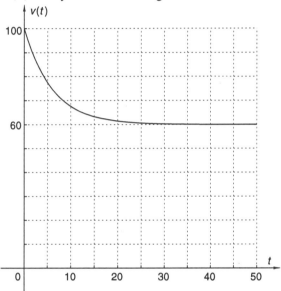

1. What does your velocity seem to be between t = 30 and t = 50 seconds? How far do you travel in the time interval [30, 50]?

2. Explain why the answer to Problem 1 can be represented as the area of a *rectangular* region of the graph. Shade this region.

3. The distance you travel between $t = 0$ and $t = 20$ can also be represented as the area of a region bounded by the (curved) graph. Count the number of squares in this region. Estimate the area of parts of squares to the nearest 0.1 square space. For instance, how would you count this partial square?

4. How many feet does each small square on the graph represent? How far, therefore, did you go in the time interval [0, 20]?

5. Problems 3 and 4 involve finding the product of the x-value and the y-value for a function where y may *vary* with x. Such a product is called the **definite integral** of y with respect to x. Based on the units of t and v(t), explain why the definite integral of v(t) with respect to t in Problem 4 has feet for its units.

6. The graph shows the cross-sectional area, y square inches, of a football as a function of the distance, x inches, from one of its ends. Estimate the definite integral of y with respect to x.

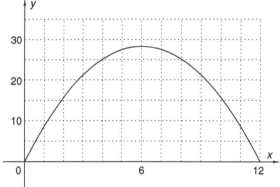

7. What are the units of the definite integral in Problem 6? What, therefore, do you suppose the definite integral represents?

8. What did you learn as a result of doing this Exploration that you did not know before? (Over)

Exploration 4: Definite Integrals by Trapezoidal Rule

<u>Objective</u>: Estimate the definite integral of a function numerically rather than graphically by counting squares.

<u>Rocket Problem</u>: Ella Vader (Darth's daughter) is driving in her rocket ship. At time $t = 0$ minutes she fires her rocket engine. The ship speeds up for a while, then slows down as Alderaan's gravity takes its effect. The graph of her velocity, $v(t)$ miles per minute, is shown below.

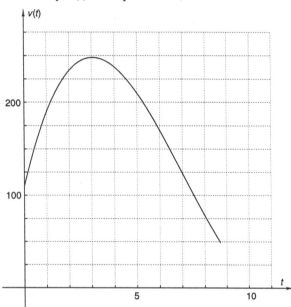

1. What mathematical concept would be used to estimate the distance Ella goes between $t = 0$ and $t = 8$?

2. Estimate the distance in Problem 1 geometrically.

3. Ella figures that her velocity is given by

   $v(t) = t^3 - 21t^2 + 100t + 110.$

   Plot this graph on your grapher. Does the graph confirm or refute what Ella figures? Tell how you arrive at your conclusion.

4. Divide the region under the graph from $t = 0$ to $t = 8$, which represents the distance, into four vertical strips of equal width. Draw four trapezoids whose areas approximate the areas of these strips, and whose parallel sides extend from the x-axis to the graph. By finding the areas of these trapezoids, estimate the distance Ella goes. Does the answer agree with Problem 2?

5. The technique in Problem 4 is the **trapezoidal rule.** Put a program into your grapher to use this rule. The function equation may be stored as y_1. The input should be the starting time, the ending time, and the number of trapezoids. The output should be the value of the definite integral. Test your program by using it to answer Problem 4.

6. Use the program from Problem 5 to estimate the definite integral using 20 trapezoids.

7. The *exact* value of the definite integral is the *limit* of the estimates by trapezoids as the width of each trapezoid approaches zero. By using the program from Problem 5, make a conjecture about the exact value of the definite integral.

8. What is the fastest Ella went? At what time was that?

9. Approximately what was Ella's rate of change of velocity when $t = 5$? Was she speeding up or slowing down at that time?

10. At what time does Ella stop? Based on the graph, does she stop abruptly or gradually?

11. What did you learn as a result of doing this Exploration that you did not know before? (Over)

Calculus Explorations
© 1998 Key Curriculum Press

Exploration 5: Introduction to Limits

<u>Objective</u>: Find the limit of a function that approaches an indeterminate form at a particular value of *x* and relate it to the definition.

1. Plot on your grapher the graph of this function.

$$f(x) = \frac{x^3 - 7x^2 + 17x - 15}{x - 3}$$

Use a friendly window with *x* = 3 as a grid point. Sketch the results here. Show the behavior of the function in a neighborhood of *x* = 3.

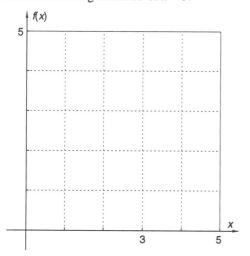

2. Substitute 3 for *x* in the equation for *f*(*x*). What form does the answer take? What name is given to an expression of this form?

3. The graph of *f* has a **removable discontinuity** at *x* = 3. The *y*-value at this discontinuity is the **limit** of *f*(*x*) as *x* approaches 3. What number does this limit equal?

4. Make a table of values of *f*(*x*) for each 0.1 unit change in *x*-value from 2.5 through 3.5.

x	*f*(*x*)
2.5	
2.6	
2.7	
2.8	
2.9	
3.0	
3.1	
3.2	
3.3	
3.4	
3.5	

5. Between what two numbers does *f*(*x*) stay when *x* is kept in the open interval (2.5, 3.5)?

6. Simplify the fraction for *f*(*x*). Solve numerically to find the two numbers close to 3 between which *x* must be kept if *f*(*x*) is to stay between 1.99 and 2.01.

7. How far from *x* = 3 (to the left and to the right) are the two *x*-values in Problem 6?

8. For the statement "If *x* is within _____ units of 3 (but not equal to 3), then *f*(*x*) is within 0.01 unit of 2," write the largest number that can go in the blank.

9. Write the definition of limit.

10. Problem 8 gives four numbers that correspond to *L*, *c*, epsilon, and delta in the definition of limit. Which is which?

11. What did you learn as a result of doing this Exploration that you did not know before? (Over)

Exploration 6: The Definition of Limit

Objective: Interpret graphically and algebraically the definition of limit.

Let *f* be the function whose graph is shown here.

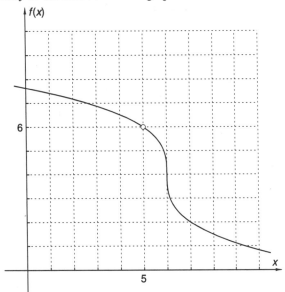

1. The limit of $f(x)$ as x approaches 5 is equal to 6. Write the definition of limit as it applies to f at this point.

2. Let $e = 1$. From the graph, estimate how close to 5 on the left side x must be kept in order for $f(x)$ to be within e units of 6.

3. From the graph, estimate how close to 5 on the right side x must be kept in order for $f(x)$ to be within $e = 1$ unit of 6.

4. For $e = 1$, approximately what must d equal in the definition of limit in order for $f(x)$ to be within e units of 6 whenever x is within d units of 5 (but not equal to 5)?

5. The equation of the function graphed is
   $$f(x) = 4 - 2(x - 6)^{1/3}, \text{ for } x \neq 5.$$
   Calculate precisely the value of d from Problem 4.

6. If $e = 0.01$, calculate precisely what d must equal in order for $f(x)$ to be within e units of 6 whenever x is within d units of 5 (but not equal to 5).

7. If $e = 0.0001$, calculate precisely what d must equal in order for $f(x)$ to be within e units of 6 whenever x is within d units of 5 ($x \neq 5$).

8. Does it appear that there is a positive value of d for *any* positive number e, no matter how small? See if you can find d algebraically in terms of e.

9. What did you learn as a result of doing this Exploration that you did not know before? (Over)

Calculus Explorations
© 1998 Key Curriculum Press

Exploration 7: Extension of the Limit Theorems by Mathematical Induction

<u>Objective</u>: Prove that the limit of a sum property is true for the sum of *any* finite number of terms.

Suppose that $f_n(x)$ is the sum of n other functions,

$$f_n(x) = g_1(x) + g_2(x) + g_3(x) + \ldots + g_n(x),$$

and that $g_1(x), g_2(x), g_3(x), \ldots, g_n(x)$ have limits $L_1, L_2, L_3, \ldots, L_n$, respectively, as x approaches c. Prove that

$$\lim_{x \to c} f_n(x) = L_1 + L_2 + L_3 + \ldots + L_n$$

for *all* integers $n \geq 2$.

<u>Proof</u>: (You supply the details!)

1. Explain how you know that the property is true for $n = 2$. (This fact is called the **anchor** of the proof.)

2. Assume that the property is false. What does this assumption tell you about values of n?

3. Let j be a value of n for which the property is not true. Let T be the set of values of n for which the property *is* true, and let F be the set of values of n for which the property is *false*. On the Venn diagram below show a value of n that is in T and a value of n that is in F.

4. By a clever use of the associative property for addition, you can turn a sum of three terms into a sum of two terms. By doing this, show that the property is true for $n = 3$. You may write "lim" as an abbreviation for limit as $x \to c$. Write 3 in T.

5. Suppose you assume that the property is true for $n = 5$. Show how you could prove that it is also true for $n = 6$.

6. The **well-ordering axiom** states that any nonempty set of positive integers has a *least* element. How do you know that set F is a nonempty set of positive integers?

7. Let l be the least element of F. How do you know that $l - 1$ is a *positive* integer? In which set is $l - 1$? Show l and $l - 1$ in the Venn diagram of Problem 3.

8. Write a statement about the limit of $f_{l-1}(x)$ and a statement about the limit of $f_l(x)$ as x approaches c.

9. By the definition of $f_n(x)$,

    $f_l(x) =$
    $g_1(x) + g_2(x) + g_3(x) + \ldots + g_{l-1}(x) + g_l(x).$

    By clever use of the associative property, as in Problem 5, associate the right side of this equation into a sum of *two* terms.

Exploration 7, continued

10. Take the limit of both sides of the equation in Problem 9. Use the anchor to write the right side of the equation as a sum of *two* limits.

The proof process in Problems 1 through 13 can be shortened if you realize that all you need to do is (a) find an anchor, and (b) show that if the property is true for some integer $n = k$, then it is also true for the next integer, $n = k + 1$. This shortened process is called **mathematical induction.** Complete the following for the property of the limit of a sum.

Proof:

Anchor:

11. Use the equation for the limit of $f_{l-1}(x)$ from Problem 8 to simplify the equation in Problem 10.

Induction hypothesis: Assume the property is true for $n = k > 2$.

Verification for $n = k + 1$:

12. Explain why the simplified equation in Problem 11 contradicts what you wrote about the limit of $f_l(x)$ in Problem 8.

13. The only place in the steps above that could account for the contradiction in Problem 12 is the assumption in Problem 2 that the property was *not* true for all integers $n \geq 2$. What can you conclude about this assumption? What can you conclude about the property?

Conclusion:

Q.E.D.

14. What did you learn as a result of doing this Exploration that you did not know before? (Over)

Exploration 8: Continuous and Discontinuous Functions

<u>Objective</u>: Given a function specified by two different rules, make the function continuous at the boundary between the two branches.

Let f be the function defined by

$$f(x) = \begin{cases} x + 1, \text{ if } x < 2 \\ k(x - 5)^2, \text{ if } x \geq 2 \end{cases}$$

where k stands for a constant.

1. Plot the graph of f for $k = 1$. Sketch the result.

2. Function f is **discontinuous** at $x = 2$. Tell what it means for a function to be discontinuous.

3. Find $\lim\limits_{x \to 2^-} f(x)$ and $\lim\limits_{x \to 2^+} f(x)$. (The second limit will be in terms of k.) What must be true of these two limits for f to be **continuous** at $x = 2$?

4. Find the value of k that makes f continuous at $x = 2$. Sketch the graph of f for this value of k.

5. The graph in Problem 4 has a **cusp** at $x = 2$. What is the origin of the word *cusp,* and why is it appropriate to use in this context?

6. Suppose someone asks, "Is $f(x)$ increasing or decreasing at $x = 2$ with k as in Problem 4?" How would you have to answer that question? What, then, can you conclude about the derivative of a function at a point where the graph has a cusp?

7. What did you learn as a result of doing this Exploration that you did not know before?

Exploration 9: Limits Involving Infinity

<u>Objective:</u> Discover what it means for a function to approach a limit as *x* approaches infinity.

A pendulum is pulled away from its rest position and let go. As it swings back and forth, its distance, $d(t)$ cm, from the wall is given by the equation

$$d(t) = 50 + 30(0.9)^t \cdot \cos \pi t,$$

where *t* is time in seconds since it was let go. The graph of function *d* shows that friction decreases the amplitude of the swings as time goes on.

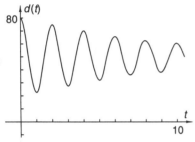

1. Plot the graph of *d* using an *x*-window of about [20, 40] and the *y*-window shown in the graph. Sketch the result here.

2. What number does $d(t)$ stay close to as *t* becomes very large?

3. The number in Problem 2 is the **limit of $d(t)$ as *t* approaches infinity.** How large does *t* have to be in order for $d(t)$ to stay within $\varepsilon = 0.1$ unit of the limit for all larger values of *t*?

4. Tell the real-world meaning of the limit in Problem 2.

5. True or false: "The larger *t* gets, the closer $d(t)$ gets to the number in Problem 2." Explain how you arrived at your answer.

6. The definition of limit states that *L* is the limit of $f(x)$ as *x* approaches *c* if and only if for any $\varepsilon > 0$ there is a number $\delta > 0$ such that if *x* is within δ units of *c* $(x \neq c)$, then $f(x)$ is within ε units of *L*. But infinity is not a number; you can't keep *x* close to infinity. You can, however, keep *x* arbitrarily far away from zero. Using the symbol ∞ to stand for infinity, tell how the definition of limit can be modified to give meaning to

$$L = \lim_{x \to \infty} f(x).$$

7. Why is the commonly used terminology ". . . $f(x)$ approaches *L* . . ." somewhat misleading? What words would better describe the relationship between $f(x)$ and *L*?

8. What did you learn as a result of doing this Exploration that you did not know before? (Over)

Calculus Explorations
© 1998 Key Curriculum Press

Exploration 10: Partial Rehearsal for Test on Limits

<u>Objective</u>: Make sure you understand the epsilon-delta definition of limit, and how to use it to calculate a derivative.

1. Let $f(x) = x^3$. Find $\lim\limits_{x \to 2} f(x)$.

   How close must x be kept to 2 in order to make $f(x)$ to stay within 0.1 unit of $f(2)$? How close must x be kept to 2 in order to make $f(x)$ to stay within e units of $f(2)$? How does your answer to this question show that $f(2)$ really *is* the limit of $f(x)$ as x approaches 2?

2. Let $g(x) = \sin \pi x$. Evaluate $g(0.5)$. Sketch the graph of g, showing at least one full cycle. How close must x be kept to 0.5 in order to make $g(x)$ stay within 0.01 unit of $g(0.5)$? Show that there is a positive number d for any $e > 0$ such that if x is within d units of 0.5 (but not equal to 0.5), then $g(x)$ is within e units of $g(0.5)$.

3. Let $h(x) = 2^{x-3} + 5$. Find $h(3)$. How close must x be kept to 3 in order to make $h(x)$ stay within e units of $h(3)$?

4. Let $r(x) = \dfrac{\sin \pi x - 0.5}{x - 1/6}$.

   What form does $r(1/6)$ take? What name is given to forms like this? See if you can figure out the limit of $r(x)$ as x approaches 1/6. See if you can figure out what key(s) to press on your calculator to find this number.

5. What did you learn as a result of doing this exercise that you did not know before?

Exploration 11: One-Problem Summary of Calculus So Far

Objective: Find approximate values of definite integrals, and approximate and exact values of derivatives.

In this problem you will operate on the function graphed here.

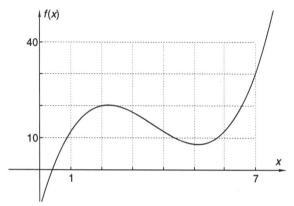

1. Estimate the definite integral of $f(x)$ from $x = 1$ to $x = 7$ by counting squares.

2. Estimate the derivative of $f(x)$ at $x = 1$ by drawing a line tangent to the graph at that point and measuring its slope.

3. The equation of f is $f(x) = x^3 - 11x^2 + 34x - 12$. Confirm on your grapher that this equation really produces the above graph. (The GRID ON feature will help.)

4. Estimate the derivative of $f(x)$ at $x = 1$ by using $f(1.01)$ and $f(0.99)$. Did the answer come out fairly close to the one you got in Problem 2?

5. Estimate the definite integral of $f(x)$ from $x = 1$ to $x = 7$ using the trapezoidal rule with $n = 6$ increments. Did the answer come fairly close to the one you got in Problem 1?

6. The exact value of the integral in Problem 5 happens to be an *integer*. By an appropriate technique, find out what that integer is.

7. Based on your technique in Problem 6, what would you propose as a definition for the definite integral of a function f between $x = a$ and $x = b$?

8. Write a difference quotient for the average rate of change of $f(x)$ from $x = 1$ to $x = x$. Plot the difference quotient on your grapher. Use a friendly window that includes $x = 1$. Sketch the graph here.

9. Do the appropriate algebra to simplify the difference quotient in Problem 8. Then find the limit of the difference quotient as x approaches 1. This answer is the *exact* value of the derivative.

10. Prove that your answer to Problem 9 really is the limit of the difference quotient by applying the appropriate limit theorems.

11. What did you learn as a result of doing this Exploration that you did not know before? (Over)

Calculus Explorations
© 1998 Key Curriculum Press

Exploration 12: Exact Value of a Derivative

Objective: Use the definition of derivative to find the exact value of the derivative of a function at a given point.

The figure shows the graph of

$$f(x) = x^3 - 4x^2 - 9x + 46.$$

In this exercise you will use the definition of derivative to find the *exact* value of $f'(4)$.

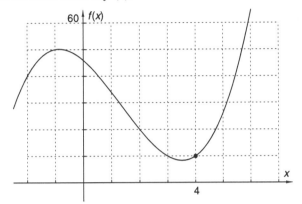

1. Find $f(4)$. Show that your answer agrees with the graph.

2. Write the definition of derivative as it applies to f at $x = 4$.

3. Substitute the values of $f(x)$ and $f(4)$ into the definition in Problem 2. Then simplify the resulting rational expression, and take the limit.

4. Plot a line on the graph at $(4, f(4))$ that has slope $f'(4)$. Observe the different scales on the two axes. Tell how the line confirms that the derivative is correct.

5. Find the exact value of $f'(2)$ using the same procedure you used for $f'(4)$. How can you tell quickly that your answer is reasonable?

6. What did you learn as a result of doing this Exploration that you did not know before?

Exploration 13: Numerical Derivative by Grapher

<u>Objective</u>: Plot the numerical derivative of a function and make connections between the derivative graph and the function graph.

In Exploration 1 you explored the function

$$d(t) = 200t \cdot 2^{-t}.$$

where t is the number of seconds since you pushed open a door and $d(t)$ is the number of degrees the door is from its rest position. The figure below is an accurate graph of function d. In this exploration you will use the numerical derivative feature of your grapher to calculate values of $d'(t)$, the instantaneous rate at which the door is opening.

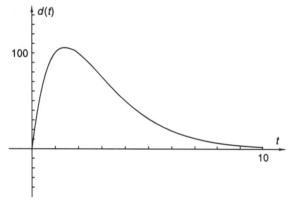

1. Confirm that the graph shown is correct by plotting the equation as y_1 on your grapher. Use a friendly x-window of about [0, 10].

2. Use a symmetric difference quotient with $\Delta t = 0.001$ to find an estimate of $d'(1)$.

3. Use the numerical derivative feature of your grapher to find an estimate of $d'(1)$. Does the answer agree with that in Problem 2?

4. Estimate $d'(2)$ numerically. In what way do the values of $d'(1)$ and $d'(2)$ correspond to the graph? What do the signs of $d'(1)$ and $d'(2)$ tell you about the motion of the door?

5. Plot the numerical derivative, $d'(t)$, as y_2. Have your instructor check your graph.   _____

6. What is true about the graph of d at the point where $d'(t) = 0$? What is happening to the door's motion at this time?

7. Use the SOLVE feature of your grapher to calculate precisely the value of t at which $d'(t) = 0$.

8. Use the MINIMUM feature of your grapher to find precisely the value of t at which $d'(t)$ is a minimum. What does $d(t)$ equal at this value of t? Put a dot at this point on the graph on this sheet.

9. The point in Problem 8 is called a **point of inflection.** Why do you suppose this name is used?

10. What did you learn as a result of doing this Exploration that you did not know before?

Calculus Explorations
© 1998 Key Curriculum Press

Exploration 14: Algebraic Derivative of a Power Function

<u>Objective</u>: Find an equation for the derivative of a power function.

The graph below shows the power function

$$f(x) = x^5.$$

In this exploration you will find a formula for $f'(c)$, the derivative at $x = c$, and from the answer figure out a way to find a formula for $f'(x)$ for *any* power function.

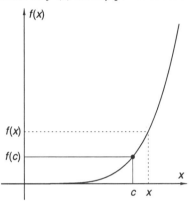

1. By the definition of derivative,

$$f'(c) = \lim_{x \to c} \frac{f(x) - f(c)}{x - c}$$

Substitute for $f(x)$ and for $f(c)$.

2. The numerator of the fraction in Problem 1 is $x^5 - c^5$. Use what you have learned in previous courses about a difference of two like powers to factor this expression. Then use the result to simplify the equation for $f'(c)$ in Problem 1, and take the limit.

3. Use the formula in Problem 2 to find the *exact* value of $f'(3)$. Confirm the answer by estimating $f'(3)$ numerically on your grapher.

4. By observing the pattern for the derivative of $f(x) = x^5$, make a conjecture about a formula for $f'(c)$ if

$$f(x) = x^{10}.$$

Test your conjecture by finding the numerical derivative for a particular value of $x = c$.

5. Based on your work in this Exploration, what seems to be a formula for $f'(x)$ if $f(x) = x^n$, where n is *any* positive integer?

6. What did you learn as a result of doing this Exploration that you did not know before?

Exploration 15: Deriving Velocity and Acceleration from Displacement Data

Objective: Find a regression equation from displacement data, and use the equation to find velocity and acceleration.

A bullet is fired from a rifle. The table shows the bullet's (horizontal) displacement from the rifle at various times after it was fired.

t seconds	x(t) feet
0.3	1000
0.6	1900
0.9	2700
1.2	3400
1.5	4000
1.8	4500
2.1	4900

1. Find the particular equation (the **regression equation**) for the best-fitting power function,

$$x(t) = at^b$$

where a and b represent constants. What does the correlation coefficient equal?

2. On your grapher, make a scatter plot of the data. On the same screen, plot the regression equation from Problem 1. Does the equation seem to fit the data reasonably well?

3. The **velocity** of the bullet is the instantaneous rate of change of displacement. Write an equation for the velocity as a function of time.

4. Use the equation in Problem 3 to calculate the velocity at $t = 0.9$ sec.

5. Confirm your answer to Problem 4 by estimating the velocity directly from the data.

6. The **acceleration** of the bullet is the instantaneous rate of change of the velocity. Write an equation for the acceleration as a function of time.

7. The equations you have derived in this Exploration are advantageous because they can be used at values of t that are *not* in the data table. Calculate the displacement, velocity, and acceleration at $t = 1$.

8. Finding a value of displacement *between* two data points is called **interpolation.** Finding a value *beyond* the data is called **extrapolation.** Suppose you use the equation in Problem 1 to extrapolate to 1 minute. Do you think the answer would match the actual displacement reasonably well? Explain.

9. What did you learn as a result of doing this Exploration that you did not know before? (Over)

Calculus Explorations
© 1998 Key Curriculum Press

Exploration 16: Derivative of the Sine of a Function

<u>Objective</u>: Find the derivative of a sine function if the argument is a *function* of x.

1. Plot the graph of $f(x) = \sin x$ on your grapher. Use a friendly window starting at $x = 0$ and ending at some convenient place around $x = 10$. Sketch the resulting graph here.

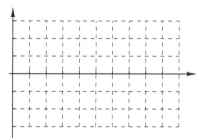

2. You have conjectured that the derivative of $f(x) = \sin x$ is $f'(x) = \cos x$. Plot $\cos x$ as y_2 and the numerical derivative of $f(x) = \sin x$ as y_3. How do the graphs confirm the conjecture?

3. Plot the graph of $g(x) = \sin 3x$ on your grapher. Sketch the resulting graph here.

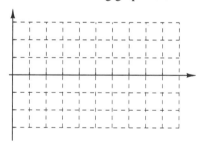

4. What is the effect on the sine graph of the constant 3 in the equation?

5. Make a conjecture about what the equation for $g'(x)$ would be. Then verify (or refute!) your conjecture by plotting $y_2 =$ your presumed equation and $y_3 =$ the numerical derivative of $g(x)$. Draw the correct g' graph on the axes in Problem 3. If your conjecture was wrong, write the correct equation for $g'(x)$.

6. Plot the graph of $h(x) = \sin x^2$ on your grapher. Use a friendly window starting at $x = 0$ and ending at some convenient place around $x = 5$. Sketch the resulting graph here.

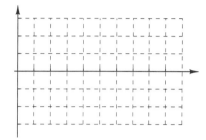

7. Make a conjecture about what an equation for $h'(x)$ might be. Then verify (or refute!) your conjecture by appropriate graphing on your grapher. Sketch the correct h' graph on the axes in Problem 6. If your conjecture was wrong, see if you can figure out a correct equation for $h'(x)$.

8. You have conjectured that the derivative of a power formula works for any constant exponent. Assuming that this conjecture is true, use the patterns you have observed in the above problems to make a conjecture about the derivative function for $t(x) = \sin x^{0.7}$. Verify your conjecture by appropriate use of the grapher. You may set the x-window back to 0 to 10 and the y-window -1 to 1. Sketch the graphs of t and t' on these axes.

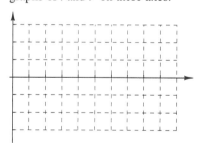

9. Suppose that $f(x) = \sin [g(x)]$ for some differentiable function g. Write a statement telling how you could find an equation for $f'(x)$.

10. What did you learn as a result of doing this Exploration that you did not know before? (Over)

Exploration 17: Rubber-Band Chain Rule Problem

Objective: Given data for a composite function, demonstrate that the chain rule gives correct answers.

Calvin pulls back a rubber band and shoots it. He figures that the force, F ounces, with which he pulls is a function of x inches, the length of the rubber band, and x is a function of t, the number of seconds since he started pulling. The following are corresponding values of t, x, and F.

t sec	x in.	F oz
0	3.0	0
0.2	4.8	4.4
0.4	6.1	8.2
0.6	6.9	11.2
0.8	7.3	13.7
1.0	7.7	14.4
1.2	7.9	15.6
1.4	8.0	16.0
1.6	8.0	16.0
1.8	3.0	0

1. Plot the graphs of F versus x and x versus t. Connect the dots with smooth curves.

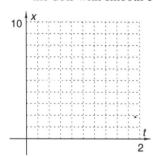 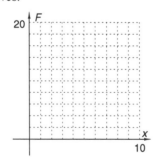

2. Estimate dx/dt at $t = 0.8$ sec. What are the units of dx/dt?

3. Estimate dF/dx at $x = 7.3$ inches (that is, when $t = 0.8$). What are the units of dF/dx?

4. Draw lines on the two graphs in Problem 1 to show geometrically that the answers to Problems 2 and 3 are correct. Observe the different scales on the axes.

5. Plot the graph of F versus t.

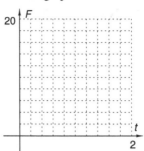

6. The **chain rule** states $\dfrac{dF}{dt} = \dfrac{dF}{dx} \cdot \dfrac{dx}{dt}$. Find an estimate of dF/dt at $t = 0.8$ sec. using the answers to Problems 2 and 3. Show how the units of dF/dx and dx/dt combine to give the units of dF/dt.

7. Find dF/dt at $t = 0.8$ sec. directly from t and F data in the table. How does the answer compare with the one you got using the chain rule?

8. How can you show geometrically that your answers to Problems 6 and 7 are correct?

9. What did you learn as a result of doing this Exploration that you did not know before?

Calculus Explorations
© 1998 Key Curriculum Press

Exploration 18: Algebraic Derivative of Sine Problem

<u>Objective</u>: Confirm algebraically that $\frac{d}{dx}(\sin x) = \cos x$, a property that was discovered geometrically.

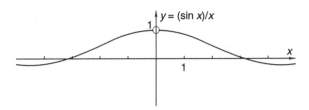

1. The graph of $y = \dfrac{\sin x}{x}$, shown above, seems to approach $y = 1$ at the discontinuity at $x = 0$. Give numerical evidence to confirm this observation.

2. The figure below shows an angle of x radians cutting off an arc x units long on a unit circle. From the figure it appears that $\sin x < x < \tan x$. Give numerical evidence to show that this is true if positive x is kept close to zero.

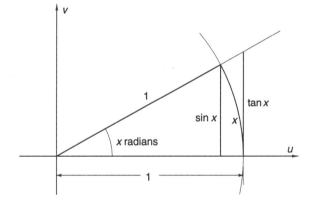

3. Transform the inequality in Problem 2 to show that for small positive values of x,

$$1 < \frac{x}{\sin x} < \sec x.$$

4. The **squeeze theorem** states that if $f(x)$ is between $g(x)$ and $h(x)$ for all x in a neighborhood of c, and $g(x)$ and $h(x)$ both approach L for a limit as x approaches c, then the limit of $f(x)$ as x approaches c is also equal to L. Use the squeeze theorem to prove that $(\sin x)/x$ approaches 1 as x approaches 0 from the right.

5. $\dfrac{d}{dx}(\sin x) = \lim\limits_{\Delta x \to 0} \dfrac{\sin (x + \Delta x) - \sin x}{\Delta x}$
   Use the appropriate properties to show that the limit equals $\cos x$.

6. The property you proved in Problem 4 can be called a **lemma** for proving the property in Problem 5. What is meant by a lemma?

7. What did you learn as a result of doing this Exploration that you did not know before? (Over)

Exploration 19: Displacement and Acceleration from Velocity

<u>Objective</u>: Given an equation for the velocity of a moving object, find equations for the displacement and for the acceleration.

Ray Sing accelerates to pass a truck. His velocity is given by

$$v(t) = 50 + 6t^{0.6},$$

where $v(t)$ is in feet per second and t is the number of seconds since he started accelerating.

1. You recall that velocity is the derivative of displacement. Let $d(t)$ be Ray's displacement from a speed limit sign on the highway. Figure out what $d(t)$ could equal so that its derivative is $50 + 6t^{0.6}$.

2. The function for $d(t)$ is called an **antiderivative** of $v(t)$, since its derivative is $v(t)$. Explain why $d(t)$ could also equal C plus your answer in Problem 1, where C stands for a constant.

3. Suppose that Ray is at a displacement of 100 feet from the speed limit sign at $t = 0$ when he first starts accelerating. Find the value of C to add to your answer in Problem 1 so that $d(t)$ will be Ray's displacement from the sign.

4. The ordered pair $(t, d(t)) = (0, 100)$ is called an **initial condition.** Why do you suppose this name is used?

5. What will Ray's displacement from the speed limit sign be when $t = 10$ seconds? 20 seconds?

6. Ray's acceleration, $a(t)$, is the derivative of his velocity. Write an equation for $a(t)$.

7. Is Ray's acceleration higher at 10 seconds or at 20 seconds? How much higher?

8. Suppose that the truck is going a constant 50 ft/sec, and is at a displacement of 160 feet from the speed limit sign when Ray starts speeding up. At what time t will Ray be 70 feet beyond the truck so that he can pull back into the proper lane?

9. What did you learn as a result of doing this Exploration that you did not know before?

Calculus Explorations
© 1998 Key Curriculum Press

Exploration 20: Derivative of a Product

<u>Objective</u>: Make a conjecture about an algebraic formula for the derivative of a product of two functions.

1. Let $g(x) = x^7$ and let $h(x) = x^{11}$.
   Let $f(x) = g(x) \cdot h(x)$.
   Find $g'(x)$ and $h'(x)$.

2. Write an equation for $f(x)$ as a single power of x.
   Then find an equation for $f'(x)$.

3. Show that $f'(x)$ does *not* equal $g'(x) \cdot h'(x)$.

4. It is possible to get the correct answer for $f'(x)$ by a clever combination of the equations for $g(x)$, $h(x)$, $g'(x)$, and $h'(x)$. For instance, you might notice that the 18 in $18x^{17}$ is the *sum* of the 7 and 11 in $7x^6$ and $11x^{10}$. Figure out what this combination is.

5. Make a conjecture about what $f'(x)$ equals in terms of $g(x)$, $h(x)$, $g'(x)$, and $h'(x)$.

6. Assume that your conjecture in Problem 5 is true for any product of two functions. If $f(x) = x^2 \sin x$, what would $f'(x)$ equal?

7. Plot on the same screen the graphs of $f(x)$, the numerical derivative of $f(x)$, and the equation for $f'(x)$ that you wrote in Problem 6. If the graphs refute your conjecture in Problem 5, change your conjecture and try again.

8. What did you learn as a result of doing this Exploration that you did not know before?

Exploration 21: Derivative of a Quotient—Do-It-Yourself!

Objective: Derive, without looking at the text, an algebraic formula for the derivative of a quotient of two functions.

1. Let $f(x) = \dfrac{x^3}{\sin x}$. Estimate $f'(1)$ numerically using your grapher.

2. Show that $f'(1)$ is definitely *not* equal to the quotient of the derivatives of the numerator and the denominator.

3. Let $y = \dfrac{u}{v}$, where u and v stand for differentiable functions of x, and where $v \neq 0$. What does $y + \Delta y$ equal in terms of u, v, Δu, and Δv?

4. Use the answer to Problem 3, and the appropriate form of the definition of derivative, to write the value of $\dfrac{dy}{dx}$

5. Transform the complex fraction from Problem 4 so that the numerator no longer contains fractions.

6. Transform the fraction in Problem 5 in such a way that the quantities $\Delta u/\Delta x$ and $\Delta v/\Delta x$ appear in an appropriate way.

7. By taking the limit in Problem 6, get a formula for $\dfrac{dy}{dx}$ that does *not* involve the limit symbol.

8. Verify that your formula in Problem 7 produces the *correct* answer for $f'(1)$ in Problem 1.

9. Write the formula in Problem 7 as a *procedure* (i.e., as a series of things you *do* to get the derivative of a quotient of two functions).

10. What did you learn as a result of doing this Exploration that you did not know before? (Over)

Calculus Explorations
© 1998 Key Curriculum Press

Exploration 22: Derivatives of Inverse Trigonometric Functions

Objective: Derive algebraic formulas for derivatives of inverse trig functions.

1. You recall that if $y = \sin^{-1} x$, then $x = \sin y$. With your grapher in parametric mode, plot the graph of $y = \sin^{-1} x$ by plotting

   $x = \sin t$
   $y = t$.

   Use equal scales on the two axes, with an x-window of about $[-10, 10]$ and a y-window of about $[-6, 6]$. Let t go from -10 to 10. Sketch the result here.

2. On the same screen plot the graph of $y = \sin x$ by using parametric mode with

   $x = t$
   $y = \sin t$.

   How are the two graphs related to each other?

3. If $y = \sin^{-1} x$, and you want to find y', you could start with the definition of derivative. However, this new problem can be transformed to an old problem by using $x = \sin y$. In this form, both sides of the equation are functions of x, but the right-hand side is a composite function, with y as the inside function. If two functions are equal, then their derivatives are equal. Use this fact to differentiate both sides of $x = \sin y$.

4. Solve the equation in Problem 3 for y' in terms of x.

5. It is possible to get the answer to Problem 4 as an *algebraic* function (involving square roots). Sketch y as an angle in standard position in a uv-coordinate system. Draw a perpendicular to the u-axis. Put x and 1 on appropriate sides of the resulting right triangle. Then write y' as an algebraic function of x.

6. Demonstrate that your formula in Problem 5 gives reasonable results by plotting a line through the point $(0.5, \sin^{-1} 0.5)$ with slope calculated by the formula.

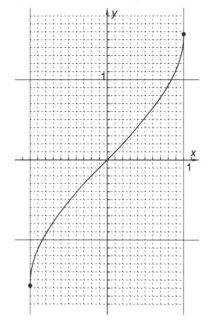

7. Derive algebraic formulas for the derivatives of $\tan^{-1} x$ and $\sec^{-1} x$. (Over)

8. What did you learn as a result of doing this Exploration that you did not know before?

Exploration 23: Differentiability Implies Continuity

<u>Objective</u>: Prove that if f is differentiable at $x = c$, then f is continuous there.

1. Write the definition of continuity at $x = c$.

2. Use the definition of continuity and the properties of limits to prove the following **lemma:**

    If $\lim\limits_{x \to c} [f(x) - f(c)] = 0$, then f is continuous at $x = c$.

3. Write the definition of derivative in the $f'(c)$ form.

4. The definition of derivative you wrote in Problem 3 contains the expression $[f(x) - f(c)]$. Why are you *not* allowed to use the limit of a quotient property to write

    $\lim\limits_{x \to c} \dfrac{f(x) - f(c)}{x - c}$

    as the limit of the numerator divided by the limit of the denominator?

5. Starting with $\lim\limits_{x \to c} [f(x) - f(c)]$, do transformations that lead to $f'(c) \cdot 0$.

6. If f is differentiable at $x = c$, then $f'(c)$ is a real number. Thus the last expression in Problem 5 is equal to zero. Explain how this fact leads to the **theorem:** "If f is differentiable at $x = c$, then f is continuous at $x = c$."

7. Give an example which shows that the converse of the theorem in Problem 6 is *false*.

8. What is meant by a *lemma* for a theorem? What word is used to describe a property that can be proved as an easy consequence of a previously proved theorem?

9. What did you learn as a result of doing this Exploration that you did not know before? (Over)

Calculus Explorations
© 1998 Key Curriculum Press

Exploration 24: Parametric Function Graphs

Objective: Analyze the motion of an object whose path is given parametrically.

1. As Adam Ant crawls along the xy-plane, his position (x, y) is given by

   $x = 0.4t \cos t$
   $y = 0.3t + 2 \sin 2t$.

   Plot the path on your grapher. Use a square window of about –5 to 5 in the x-direction and –2 to 5 in the y-direction. Use a t-range of 0 to 4π. Have your instructor check your graph.

2. Find equations for dx/dt and dy/dt. Evaluate the two derivatives at $t = 6$.

3. Use the answers to Problem 2 in an appropriate way to find the slope of the path at $t = 6$. Offer an explanation, based on the graph, why your answer is reasonable.

4. Use the CALCULATE feature of your grapher to find the answer to Problem 3 numerically.

5. Adam starts out at the origin and goes to the "northeast" for a while. At approximately what value of t does he turn around and start coming back?

6. Reset the t-range to go from 0 to the time Adam first arrives back at $y = 0$. Use a t-step of 0.01 to get a fairly accurate graph. Sketch the result here.

7. Find precisely by calculation the value of t at which y first stops increasing and starts decreasing.

8. At the time in Problem 7, is Adam stopped, or is he still moving? Justify your answer.

9. Use the ZOOM BOX feature to zoom in on the point in Problem 8. How does your answer confirm (or refute!) your conclusion in that problem?

10. What did you learn as a result of doing this Exploration that you did not know before? (Over)

Exploration 25: Implicit Relation Derivatives

<u>Objective</u>: Find the derivative of a function specified implicitly, and confirm by graphing.

The following is the graph of the implicit relation

$$x^2 - 4xy + 4y^2 + x - 12y - 10 = 0$$

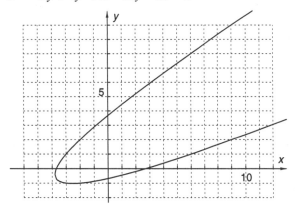

1. Tell why this relation is not a function.

2. Confirm that the graph is reasonable by calculating the values of y when $x = 6$, and showing that the corresponding points are on the graph.

3. Find an equation for y' by differentiating implicitly with respect to x. The answer will be in terms of x and y. Observe the product rule and the chain rule!

4. Use the equation from Problem 3 to calculate the values of y' at the two points you found in Problem 1. Show on the graph that the two answers are reasonable.

5. What does the formula for y' give you if you substitute $(1, 3)$ for (x, y)? Why does the answer have no meaning for this problem?

6. Find an equation for y explicitly in terms of x. Then tell why it is easier to find y' by implicit differentiation than it would be to find it directly from the explicit equation.

7. How can you tell algebraically that the graph in this exercise is a parabola?

8. What did you learn as a result of doing this Exploration that you did not know before? (Over)

Calculus Explorations
© 1998 Key Curriculum Press

Exploration 26: A Motion Antiderivative Problem

<u>Objective</u>: Given an equation for the velocity of a moving object, find equations for the displacement and acceleration.

Tay L. Gates wants to determine the characteristics of his new pickup truck. With special instruments he records its velocity at 2-second intervals as he starts off from a traffic light.

t seconds	velocity, ft/sec
0	0
2	4.5
4	6.9
6	8.8
8	10.4
10	11.9

1. Show that the power function

   $v(t) = 3t^{0.6}$

   fits these data closely.

2. Write an equation for Tay's acceleration, $a(t)$. Is the acceleration getting larger or smaller as time goes on? Tell how you figured this out.

3. Velocity is the derivative of displacement. Thus, displacement is the **antiderivative** of velocity. Write an equation for $x(t)$, the truck's displacement from the middle of the intersection. Use the fact that the truck was initially at $x(0) = -50$ feet from the center of the intersection at $t = 0$ when the light turned green.

4. Use the equation in Problem 3 to predict where Tay's truck was 10 seconds after he started accelerating.

5. Calculate the answer to Problem 4 directly from the data in the table, using the trapezoidal rule.

6. How long does it take before the truck is 100 feet beyond the intersection?

7. What did you learn as a result of doing this Exploration that you did not know before?

Exploration 27: Differentials, and Linearization of a Function

<u>Objective</u>: Find the equation of the linear function that best fits the graph of a given function at a given point.

The graph below shows the function

$$f(x) = \sec x.$$

In this Exploration you will find the equation of the linear function that best fits the graph of f when x is kept close to 1.

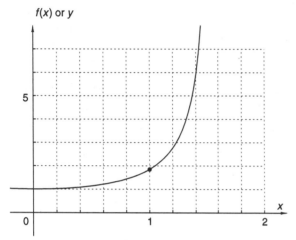

1. You recall the point-slope form, $y - y_0 = m(x - x_0)$, of the linear function equation. In this equation, m is the slope and (x_0, y_0) is a point on the graph. If the linear function is to fit function f when x is close to 1, then m should equal the slope of f at $x = 1$, and y_0 should equal $f(1)$. Thus the best-fitting linear function has an equation of the form

$$y = f(1) + f'(1)(x - 1).$$

Find the particular equation of the best-fitting linear function.

2. Plot the graph of f and the graph of the linear function on the same screen. Also, plot the linear function on the graph above. Is the linear function tangent to the graph of f at $x = 1$?

3. The quantity $f(x) - y$ is the **error** involved in using y in Problem 1 as an approximation for $f(x)$. On your grapher, make a table of values of the error for each 0.01 unit of x starting at $x = 0.95$. What happens to the error as x gets closer to 1?

4. How close to 1 should x be kept in order for the error in Problem 3 to be less than 0.001?

5. The quantity $(x - 1)$ in Problem 1 is called the **differential of x** and is abbreviated dx. The quantity $f'(1)(x - 1)$ is called the **differential of y,** abbreviated dy. Let $dx = 0.4$. Draw lines on the graph showing what dx and dy mean.

6. What "interesting" quantity does the ratio $dy \div dx$ equal?

7. Tell how dx and dy are related to Δx and Δy, which represent the changes in x and $f(x)$ as x goes from 1 to 1.4.

8. What did you learn as a result of doing this Exploration that you did not know before? (Over)

Calculus Explorations
© 1998 Key Curriculum Press

Exploration 28: Riemann Sums for Definite Integrals

<u>Objective</u>: Find an alternative to the trapezoidal rule for estimating a definite integral.

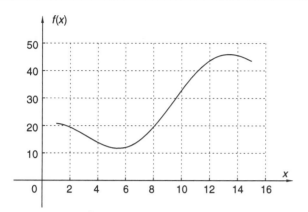

1. The symbol $\int_a^b f(x)\, dx$ stands for the definite integral of $f(x)$ from $x = a$ to $x = b$. By counting squares, find an approximation of $\int_2^{14} f(x)\, dx$ for the function graphed above.

2. Find an estimate of $f(c)$ for these values of $x = c$.

c	f(c)
3	_____
5	_____
7	_____
9	_____
11	_____
13	_____

3. On the figure above, draw rectangles between vertical grid lines with altitudes equal to $f(c)$, starting at $x = 2$ and ending at $x = 14$. Tell why the sum of the areas of these rectangles is an estimate of $\int_2^{14} f(x)\, dx$.

4. The sum of the areas of the rectangles mentioned in Problem 3 is called a **midpoint Riemann sum**. Evaluate this sum. How close does the answer come to the estimate of $\int_2^{14} f(x)\, dx$ you got by counting squares in Problem 1?

5. Find an estimate of $f(c)$ for these values of c.

c	f(c)
2	_____
4	_____
6	_____
8	_____
10	_____
12	_____

6. The values of c in Problems 2 and 5 are called **sample points** for a Riemann sum. Find an estimate of $\int_2^{14} f(x)\, dx$ using the sample points in Problem 5.

7. Tell how estimating an integral using a Riemann sum is related to doing it by trapezoidal rule.

8. What did you learn as a result of doing this Exploration that you did not know before? (Over)

Exploration 29: The Mean Value Theorem

Objective: Without looking at the text, discover the hypotheses and conclusion of the mean value theorem.

1. For $f(x) = -0.1x^3 + 1.2x^2 - 3.6x + 5$, graphed below, there is a value of $x = c$ between 3 and 7 at which the tangent to the graph is parallel to the secant line connecting $(3, f(3))$ and $(7, f(7))$. Draw the secant line and the tangent line. Approximately what is the value of c?

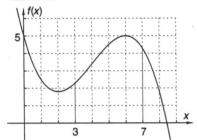

2. Function f in Problem 1 has *two* values of $x = c$ between $x = 1$ and $x = 7$ at which the tangent lines are parallel to the corresponding secant line. Draw these tangents on the graph below. Approximately what are the values of c?

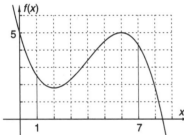

3. Function $g(x) = 6 - 2(x - 4)^{2/3}$, graphed below, is not differentiable at $x = 4$. Is there a value of $x = c$ between $x = 1$ and $x = 5$ at which the slope of the tangent line equals the slope of the corresponding secant line? If so, draw it. If not, tell why not.

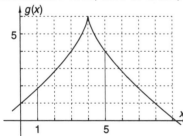

4. Function g in Problem 3 *does* have a value of $x = c$ between $x = 1$ and $x = 4$ where the tangent line parallels the corresponding secant line. This is true because the point at which the function is not differentiable occurs at the *endpoint* of the interval. Illustrate this fact on the graph in the next column. Approximately what is the value of c?

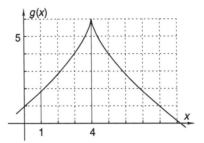

5. Function $h(x) = 0.2(x - 2)^2 - \dfrac{|x - 5|}{x - 5}$ if $x \neq 5$, and $h(5) = 2.8$. Is h differentiable for all x in the *open* interval $(5, 8)$? Is h continuous at $x = 5$? Is there a value of $x = c$ in $(5, 8)$ for which $h'(c)$ equals the slope of the secant line connecting $(5, h(5))$ and $(8, h(8))$? Illustrate your answer.

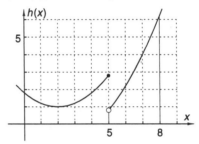

6. Function h in Problem 5 is differentiable on $(1, 5)$ and discontinuous at $x = 5$. Is there a point $x = c$ in the interval $(1, 5)$ at which $h'(c)$ equals the slope of the corresponding secant line? Illustrate your answer on the graph below.

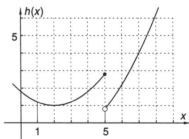

7. The number $x = c$ in the above problems is the "mean" value referred to in the mean value theorem. State the mean value theorem. Explain why the hypotheses are **sufficient** conditions but *not* **necessary** conditions.

Exploration 30: Some Very Special Riemann Sums

<u>Objective</u>: Calculate Riemann sums for given sets of sample points and reach a conclusion about how the sample points were chosen.

The figure shows the graph of $f(x) = x^{1/2}$. In this Exploration you will integrate $f(x)$ from $x = 1$ to 4.

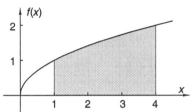

1. Find an estimate for the integral

$$I = \int_1^4 x^{1/2}\, dx$$

by trapezoidal rule with $n = 3$ subintervals. Write down all the decimal places your calculator will give you. Does this value overestimate or underestimate the actual integral? Explain.

2. Find a midpoint Riemann sum for integral I in Problem 1. Use $n = 3$ increments. Show that this sum is *not* equal to the value by trapezoidal rule. Does the midpoint sum overestimate or underestimate the actual integral? Explain.

3. Find a Riemann sum for integral I using the subintervals in Problem 1, but using the following sample points. (k stands for the subinterval number.)

k	$x = c$
1	1.48584256
2	2.49161026
3	3.49402722

How does this sum compare with the answers to Problem 1 and Problem 2?

4. Find a Riemann sum for I using six subintervals of equal width, and these sample points:

k	$x = c$
1	1.24580513
2	1.74701361
3	2.24768040
4	2.74810345
5	3.24839587
6	3.74861006

How does the integral by this Riemann sum compare with other values in this problem set?

5. Let $g(x) = \frac{2}{3}x^{3/2}$. Find the point in the open interval (1, 1.5) at which the conclusion of the mean value theorem is true for function g. Where have you seen this number in this Exploration?

6. How is function g related to function f?

7. Make a conjecture about the *exact* value of the integral in Problem 1.

8. What did you learn as a result of doing this Exploration that you did not know before? (Over)

Exploration 31: The Fundamental Theorem of Calculus

<u>Objective</u>: Prove that a definite integral can be calculated *exactly*, using an indefinite integral.

1. Write the definition of $\int_a^b f(x)\, dx$.

2. Write the definition of $g(x) = \int f(x)\, dx$.

3. How can you be sure that the mean value theorem applies to function g?

4. The figure shows function g in Problem 2. Write the conclusion of the mean value theorem as it applies to g on the interval from $x = a$ to $x = x_1$ and illustrate the conclusion on the graph.

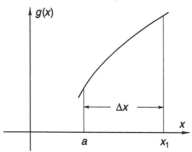

5. The figure in the next column shows the graph of $f(x)$ from Problem 2. Let $c_1, c_2, c_3, \ldots, c_n$ be sample points determined by the mean value theorem as in Problem 4. Write a Riemann sum R_n for $\int_a^b f(x)\, dx$ using these sample points and equal Δx values. Show the Riemann sum on the graph.

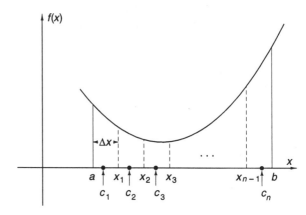

6. By the mean value theorem, $g'(c_1) = \dfrac{g(x_1) - g(a)}{\Delta x}$, and so on. By the definition of indefinite integral, $g'(c_1) = f(c_1)$. By appropriate substitutions, show that R_n from Problem 5 is equal to $g(b) - g(a)$.

7. R_n from Problem 6 is *independent* of n, the number of increments. Use this fact, and the fact that $L_n \leq R_n \leq U_n$ to prove that

$$\int_a^b f(x)\, dx = g(b) - g(a).$$

8. The conclusion in Problem 7 is called the **fundamental theorem of calculus.** Show that you understand what it says by using it to find the *exact* value of $\int_1^4 x^{1/2}\, dx$. (Over)

Exploration 32: Some Properties of Definite Integrals

Objective: Illustrate by graph or by the fundamental theorem of calculus that certain properties of definite integrals are true.

1. The graph shows $f(x) = 19 - x^2$. Tell why f is called an **even function.**

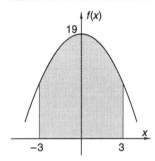

2. Use the fundamental theorem to calculate

$$\int_{-3}^{3} (19 - x^2)\, dx \text{ and } \int_{0}^{3} (19 - x^2)\, dx.$$

Explain why the first integral is *twice* the second.

3. The graph shows $g(x) = x^3$. Explain why g is called an **odd function.**

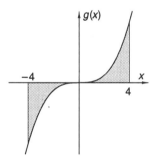

4. Use the fundamental theorem to calculate $\int_{-4}^{4} x^3\, dx$.
   Explain why the answer equals zero.

5. The graph shows the function $h(x) = 0.4x^3 - 3x^2 + 5x + 5$.

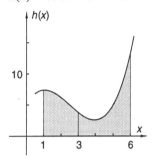

Use the fundamental theorem to calculate

$$\int_{1}^{3} h(x)\, dx \text{ and } \int_{3}^{6} h(x)\, dx.$$

6. Quick! Calculate $\int_{1}^{6} h(x)\, dx$.

7. Given:

$$\int_{2}^{7} u(x)\, dx = 29, \int_{2}^{7} v(x)\, dx = 13, \int_{2}^{4} v(x)\, dx = 8$$

a. Find $\int_{2}^{7} (u(x) + v(x))\, dx$.

b. Find $\int_{4}^{7} v(x)\, dx$.

8. What did you learn as a result of doing this Exploration that you did not know before? (Over)

Exploration 33: Applications of Definite Integrals

Objective: Without looking at the text, learn an interpretation of $f(x)\ dx$ in a definite integral.

Spaceship Velocity and Displacement Problem: A spaceship is fired into orbit. As the last stage of the booster rocket is fired the spaceship is going 3000 feet per second (a bit under 2000 mph). Its velocity, $v(t)$, at t seconds since the booster was fired is given by

$$v(t) = 3000 + 18t^{1.4}.$$

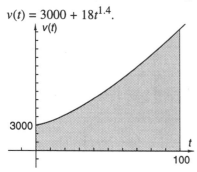

1. On the graph, pick a sample point, t, somewhere on the t-axis between 0 and 100. Show the corresponding point $(t, v(t))$ on the graph.

2. Draw a narrow vertical strip of the region in such a way that the ordered pair $(t, v(t))$ is within the strip. Label the width of the strip dt.

3. If the strip is narrow (i.e., dt is small), the velocity throughout the time interval dt is not much different from the velocity $v(t)$ at the sample point. Explain why the distance traveled in this time interval is approximately equal to $v(t) \cdot dt$.

4. Write a Riemann sum that represents, approximately, the total distance the spaceship goes between $t = 0$ and $t = 100$.

5. Explain why the Riemann sum in Problem 4 is between the corresponding lower sum and upper sum. Based on this fact, why can you conclude that the limit of the Riemann sum is *exactly* equal to $\int_0^{100} v(t)\ dt$?

6. Find the distance traveled by the spaceship by evaluating the integral in Problem 5 using the fundamental theorem of calculus.

7. In order to orbit, the spaceship must be going at least 26,000 feet per second (about 17,500 mph). To the nearest second, at what time is it going that fast?

8. How far does the spaceship go from the time the last stage fires till it reaches orbital velocity?

9. Based on what you observed while doing this exercise, why do you suppose integrals are written in *differential* form, $\int f(x)\ dx$, instead of simply in derivative form, $\int f(x)$?

Calculus Explorations
© 1998 Key Curriculum Press

Exploration 34: Derivation of Simpson's Rule

<u>Objective</u>: Derive a way to approximate definite integrals by fitting parabolas to the data.

1. Sketch a graph of $f(x) = ax^2 + bx + c$ from $x = -h$ to $x = h$. (Do *not* put the vertex at the origin.)

2. Let $y_0 = f(-h)$, $y_1 = f(0)$, and $y_2 = f(h)$. By doing the integrating, show that the area of the region under the graph from $x = -h$ to $x = h$ is
$$\text{Area} = \frac{1}{3}h(2ah^2 + 6c).$$

3. Write y_0, y_1, and y_2 in terms of a, b, c, and h. Then write $y_0 + y_2$ in terms of a, b, c, and h.

4. By appropriate algebra, show that the area of the region under the parabola may be found from the y-values and h alone, namely,
$$\text{Area} = \frac{1}{3}h(y_0 + 4y_1 + y_2).$$

5. Sketch the graph of $y = 50 \cdot 2^x$ from $x = 0$ to $x = 12$. Partition the interval $[0, 12]$ into six subintervals of width $h = 2$. Then use the result of Problem 4 to estimate
$$\int_0^{12} 50 \cdot 2^x \, dx$$
by using three parabolas, one for $[0, 4]$, one for $[4, 8]$, and one for $[8, 12]$.

6. The technique you used in Problem 5 is called **Simpson's rule.** Suppose that $\int_a^b y \, dx$ is to be evaluated using Simpson's rule with n equal increments, each of width h. Write a formula for the integral in terms of h and the values of y_i.

7. Program your grapher to evaluate integrals by Simpson's rule. Evaluate the integral in Problem 5 using $n = 100$ increments. How does the answer compare with the exact answer, $295{,}391.809\ldots$?

8. Evaluate the integral in Problem 5 using the trapezoidal rule and using midpoint Riemann sums, each with $n = 100$. Which of the three approximations is closest to the actual value, and which is farthest away?

9. Why must n be an *even* number for Simpson's rule to be used?

Exploration 35: Another Form of the Fundamental Theorem

<u>Objective</u>: Find the derivative of a definite integral from a fixed lower limit to a variable upper limit.

1. Let $f(x) = \int_1^x t^{1/2}\, dt$. Evaluate $f(9)$.

2. Sketch the graph of $y = t^{1/2}$. Show on the graph the geometrical meaning of $f(x)$.

3. Do the integrating to find an equation for $f(x)$ that does not involve the integral sign.

4. Differentiate both sides of the equation in Problem 3 to find an equation for $f'(x)$. Then tell how you could get $f'(x)$ quickly, in one step, just by looking at the definition of $f(x)$ in Problem 1.

5. Let $g(x) = \int_3^x t^3\, dt$. Quick! Find $g'(x)$.

6. Let $h(x) = \int_2^{x^3} \cos t\, dt$. Do the integrating to find an equation for $h(x)$ that does not involve the integral sign.

7. Find $h'(x)$ using your answer in Problem 6. Observe the chain rule!

8. By observing the pattern in Problems 6 and 7, evaluate the following derivative *quickly*.

$$\frac{d}{dx}\left(\int_0^{\sin x} \tan^3 (t^5)\, dt\right)$$

9. As a result of your work in Problems 1 through 8, you should be able to understand the **fundamental theorem of calculus in the derivative of an integral form,** specifically:

If $g(x) = \int_a^x f(t)\, dt$, where a stands for a constant, then $g'(x) = f(x)$.

Use this theorem to find $L'(x)$ quickly if

$$L(x) = \int_1^x \frac{1}{t}\, dt.$$

10. Explain why $L(x)$ in Problem 9 cannot be evaluated directly by the fundamental theorem in the $g(b) - g(a)$ form using the antidifferentiation methods you know so far.

Calculus Explorations
© 1998 Key Curriculum Press

Exploration 36: Natural Logs and the Uniqueness Theorem

<u>Objective:</u> Prove algebraically that ln has the properties of logarithms.

1. Write the definition of ln x.

2. Let $f(x) = \ln 3x$.
   Let $g(x) = \ln 3 + \ln x$.
   Show numerically that $f(7) = g(7)$.

3. Show *algebraically* that $f(1) = g(1)$.

4. Show that f and g meet the differentiability hypothesis of the mean value theorem on any interval $(1, b)$ where $b > 0$.

5. Tell why f and g in Problem 4 meet the continuity hypothesis of the mean value theorem at 1 and b.

6. Let $h(x) = f(x) - g(x)$. Explain why h meets the hypotheses of the mean value theorem on $[1, b]$.

7. What does $h(1)$ equal? What does the mean value theorem allow you to conclude about function h on the interval $(1, b)$?

8. Assume that there is a number $x = b > 0$ such that ln $3b \neq \ln 3 + \ln b$. Let c be the value of x in the conclusion of the mean value theorem for function h on the interval $(1, b)$. What does this assumption tell you about $h'(c)$?

9. In Problem 4 you should have found $f'(x)$ and $g'(x)$. What does $h'(x)$ equal for all $x > 0$? What, specifically, does $h'(c)$ equal? What can you say about the assumption made in Problem 8? As a result, what must be true about ln $3x$ and ln $3 + \ln x$ for *all* $x > 0$?

10. Problems 2 through 9 constitute a proof by example of the **uniqueness theorem for derivatives.** Write the statement of this theorem.

11. What did you learn as a result of doing this Exploration that you did not know before?

Exploration 37: Properties of Logarithms

<u>Objective</u>: Use the uniqueness theorem for derivatives to show that ln has the properties of logarithms.

1. Let $f(x) = \ln 7x$.
   Let $g(x) = \ln 7 + \ln x$.
   Prove that $f(x) = g(x)$ for all $x > 0$.

2. Let $p(x) = \ln 7 - \ln x$.
   Let $q(x) = \ln \dfrac{7}{x}$.
   Prove that $p(x) = q(x)$ for all $x > 0$.

3. Let $h(x) = \ln (x^5)$.
   Let $j(x) = 5 \ln x$.
   Prove that $h(x) = j(x)$ for all $x > 0$.

4. You recall the definition of logarithm from previous mathematics courses. Show that you know this definition by using it to write this equation in *exponential* form:

   $$r = \log_s t$$

5. Use the definition of logarithm to write this equation in *logarithmic* form:

   $$m^p = w$$

6. Use the definition of logarithm in an appropriate manner to derive the change-of-base property:

   $$\log_b x = \frac{\log_a x}{\log_a b} \,.$$

7. If ln really is a logarithm, it must have some number for its base. Let e be this number. That is, let

   $$\log_e x = \ln x.$$

   Use the change-of-base property to find a relationship between base e logs and base 10 logs. Then use your calculator in an appropriate manner to find out what e must equal.

8. What did you learn as a result of doing this Exploration that you did not know before? (Over)

Calculus Explorations
© 1998 Key Curriculum Press

Exploration 38: Derivative of an Exponential Function

<u>Objective</u>: Use implicit differentiation to find the derivative of an exponential function.

1. Let $y = 5^x$. Find the numerical derivative of y if $x = 2$.

2. Show that the power rule for derivatives does *not* give the correct answer for y' in Problem 1.

3. Take the ln of both sides of the equation in Problem 1. Differentiate both sides of this implicit relation, observing the chain rule. Do the necessary algebra to solve the resulting equation for y' explicitly in terms of x.

4. Does your equation in Problem 3 give the correct answer for y' if $x = 2$? If not, go back and correct your work in Problem 3.

5. Generalize the pattern in Problem 3 to find a formula for $f'(x)$ if $f(x) = b^x$, where b stands for a positive constant.

6. The figure below shows an accurate graph of

   $f(x) = 5(0.6^x)$.

   Find $f'(1)$. Demonstrate geometrically on the graph that your answer is correct.

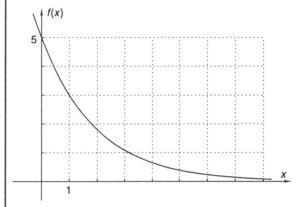

7. By observing the pattern for the derivative of an exponential function, you should be able to write the antiderivative. Find

   $\int 5^x \, dx$.

8. What did you learn as a result of doing this Exploration that you did not know before?

Exploration 39: Base *e* Logs vs. Natural Logs

<u>Objective</u>: Show that the base *e* logarithm function is identical to the natural logarithm function.

1. Let $y = (1 + 1/x)^x$. Plot the graph of y in a friendly *x*-window of about [0, 10]. Sketch the result here.

2. If you plot y from Problem 1 in a (friendly) window of about [0, 10,000], the graph looks like a horizontal straight line! Give evidence to show that the line is not really horizontal.

3. Find the number *e* by pressing e^1 on your calculator. How does the answer compare to the values of y in your graph from Problem 2? What, then, do you conjecture is the limit of $(1 + 1/x)^x$ as x approaches infinity?

4. From the text you learn that if $f(x) = \log_b x$, then $f'(x) = (1/x) \log_b e$. If the logarithms have $b = e$ as the base, what does $f'(x)$ equal?

5. Let $f(x) = \log_e x$. Let $g(x) = \ln x$. Use the uniqueness theorem for derivatives to prove that $\log_e$ and ln are identical functions.

6. If $y = \log_e x$, then the inverse function has the equation $x = \log_e y$. By using the definition of logarithm from algebra, find an equation for y in terms of x for the inverse function.

7. If $f(x) = \ln x$, then $f^{-1}(x) =$ _____. What expression goes in the blank?

8. What did you learn as a result of doing this Exploration that you did not know before?

Calculus Explorations
© 1998 Key Curriculum Press

Exploration 40: A Compound Interest Problem

<u>Objective</u>: Use a base *e* exponential function as a mathematical model of the money in a savings account as a function of time.

When Max de Monet started high school his parents invested $10,000 in a savings account to help pay for his college education. Exactly 4 years later when he graduated the account had $15,528.08 in it. Max knows that since the account pays interest compounded continuously, the general equation for $M(t)$, the number of dollars t years after the investment, is

$$M(t) = ae^{kt},$$

where a and k stand for constants and e is the base of natural logarithms.

1. By substituting the two given ordered pairs into the general equation, find the particular values of a and k for Max's account.

2. At what instantaneous rate is the amount in Max's account changing when the money was first invested at $t = 0$? At what rate was it changing when he graduated at $t = 4$?

3. Find the rate of change of Max's money as a percentage of the amount in the account at $t = 0$ and at $t = 4$. What do you notice about the two answers?

4. Show algebraically that if $M(t) = ae^{kt}$, then

$$\frac{\frac{d}{dt}\left(M(t)\right)}{M(t)}$$

is a constant. Where does this constant appear in the general equation?

5. Suppose that Max gets a scholarship, and is able to leave the money in the account until he graduates from college at $t = 8$ years. How much money would be there at that time?

6. Max anticipates getting a good job after college, and wants to keep the money in the savings account until he retires at time $t = 50$ years. How much money would he have then? Surprising?!

7. How long does it take for Max's money to double what it was when it was first invested?

8. What did you learn as a result of doing this Exploration that you did not know before? (Over)

Exploration 41: A Limit by l'Hospital's Rule

<u>Objective</u>: Find the number that dy/dx approaches if both dy/dt and dx/dt approach zero at a particular value of t.

The figure shows the hypocycloid of three cusps (the "deltoid") with parametric equations

$$x = 2 \cos t - \cos 2t$$
$$y = 2 \sin t + \sin 2t.$$

In this Exploration you will investigate the limit which dy/dx approaches as t approaches a value at one of the cusps.

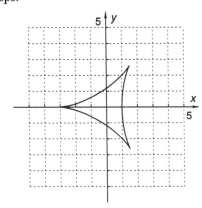

1. Show that the cusp in the first quadrant corresponds to $t = \pi/3$.

2. Find the (algebraic) derivatives dx/dt, dy/dt, and dy/dx.

3. Evaluate dx/dt and dy/dt when $t = \pi/3$. Tell why the parametric chain rule fails to give a value for dy/dx at this point.

4. Make a table of values of dy/dx starting at t equals exactly $\pi/3$ and using a small value of Δt such as 0.001. You should find that there is no value of dy/dx at $t = \pi/3$. What number does dy/dx seem to be approaching as t approaches $\pi/3$?

5. Plot a line through the cusp in Quadrant I with slope equal to the limit of dy/dx you found in Problem 4. How does the line seem to relate to the graph?

6. **L'Hospital's rule** states that if $f(x)$ and $g(x)$ both approach 0 as x approaches c, then

$$\lim_{x \to c} \frac{f(x)}{g(x)} = \lim_{x \to c} \frac{f'(x)}{g'(x)}.$$

Differentiate the numerator and denominator for dy/dx with respect to t. Show that your conjecture for the limit of dy/dx in Problem 4 agrees with the value you get by l'Hospital's rule.

7. Show that l'Hospital's rule applies to

$$\lim_{x \to 0} \frac{1 - 5x - \cos 3x}{x}.$$

Use the rule to find the limit.

8. What did you learn as a result of doing this Exploration that you did not know before? (Over)

Calculus Explorations
© 1998 Key Curriculum Press

Name _____ Group _____ Date _____

Exploration 42: Differential Equation for Compound Interest

Objective: Write and solve a differential equation for the amount of money in a savings account as a function of time.

When money is left in a savings account, it earns interest equal to a certain percent of what is there. The more money you have there, the faster it grows. If the interest is *compounded continuously,* the interest is added to the account the instant it is earned.

1. For continuously compounded interest, the instantaneous rate of change of money is directly proportional to the amount of money. Define variables for time and money, and write a **differential equation** expressing this fact.

2. **Separate the variables** in the differential equation in Problem 1, then integrate both sides with respect to t. Transform the integrated equation so that the amount of money is expressed explicitly in terms of time.

3. The integrated equation from Problem 2 will contain e raised to a power containing *two* terms. Write this power as a product of two different powers of e, one that contains the time variable, and one that contains no variable.

4. You should have the expression e^C in your answer to Problem 3. Explain why e^C is always positive.

5. Replace e^C with a new constant, C_1. If C_1 is allowed to be positive or negative, explain why you no longer need the $\pm$ sign that appears when you removed the absolute value in Problem 2.

6. Suppose that the amount of money is $1000 when time equals zero. Use this **initial condition** to evaluate C_1.

7. If the interest rate is 5% per year, then $d(\text{money})/d(\text{time}) = 0.05(\text{money})$ in dollars per year. Evaluate the proportionality constant in Problem 1.

8. How much money will be in the account after 1 year? 5 years? 10 years? 50 years? 100 years? Do the computations in the most time-efficient manner.

9. How long would it take for the amount of money to double its initial value?

10. What did you learn as a result of doing this Exploration that you did not know before? (Over)

Exploration 43: Differential Equation for Memory Retention

<u>Objective</u>: Write and solve a differential equation for the number of names remembered as a function of time.

Ira Member is a freshman at a large university. One evening he attends a reception at which there are many members of his class whom he has not met. He wants to predict how many new names he will remember at the end of the reception.

1. Ira assumes that he meets people at a constant rate of R people per hour. Unfortunately, he forgets names at a rate proportional to y, the number he remembers. The more he remembers, the faster he forgets! Let t be the number of hours he has been at the reception. What does dy/dt equal? (Use the letter k for the proportionality constant.)

2. The equation in Problem 1 is a **differential equation** since it has differentials in it. By algebra, separate the variables so that all terms containing y appear on one side of the equation and all terms containing t appear on the other side.

3. Integrate both sides of the equation in Problem 2. You should be able to make the integral of the reciprocal function appear on the side containing y.

4. Show that the solution in Problem 3 can be transformed to the form

   $ky = R - Ce^{-kt}$,

   where C is a constant related to the constant of integration. Explain what happens to the absolute value sign that you got from integrating the reciprocal function.

5. Use the initial condition $y = 0$ when $t = 0$ to evaluate the constant C.

6. Suppose that Ira meets 100 people per hour, and that he forgets at a rate of 4 names per hour when $y = 10$ names. Write the particular equation expressing y in terms of t.

7. How many names will Ira have remembered at the end of the reception, $t = 3$ hours?

8. What did you learn as a result of doing this Exploration that you did not know before?

Exploration 44: Introduction to Slope Fields

<u>Objective</u>: Find graphically a particular solution of a given differential equation, and confirm it algebraically.

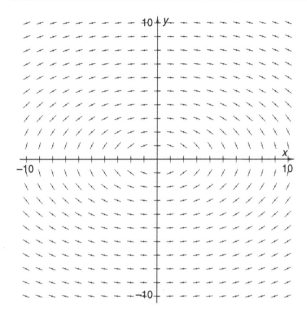

The diagram above shows the **slope field** for the differential equation

$$\frac{dy}{dx} = -\frac{0.36x}{y}.$$

1. From the differential equation, find the slope at the points (5, 2) and (–8, 9). Mark these points on the diagram. Tell why the slopes are reasonable.

2. Start at the point (0, 6). Draw a graph representing the particular solution of the differential equation which contains that point. The graph should be "parallel" to the slope lines, and be some sort of average of the slopes if it goes between lines. Go both to the right and to the left. Where does the graph seem to go after it touches the *x*-axis? What geometrical figure does the graph seem to be?

3. Start at the point (5, 2) from Problem 1 and draw another particular solution of the differential equation. How is this solution related to the one in Problem 2?

4. Solve the differential equation algebraically. Find the particular solution that contains (0, 6). Verify that the graph really *is* the figure you named in Problem 2.

5. What did you learn as a result of doing this Exploration that you did not know before?

Exploration 45: Euler's Method

<u>Objective</u>: Given a differential equation, find an approximation to a particular solution by a numerical method.

1. For the differential equation

$$\frac{dy}{dx} = -\frac{x}{2y},$$

calculate the slope at the point (0, 3). Then calculate dy if $dx = 0.5$. Use the result to estimate the value of y at $x = 0.5$.

2. Calculate the slope at $(0.5, y)$ from Problem 1. Then calculate dy if $dx = 0.5$. Use the result to estimate the value of y at $x = 1$.

3. Repeat the computations in Problems 1 and 2 for values of x from 1.5 through 7. Record the values in the table. This technique is called **Euler's method** for solving differential equations.

x	y	Slope	dy
0	3	0	0
0.5	3	–0.0833. . .	–0.0416. . .
1	2.9583. . .		
1.5			
2			
2.5			
3			
3.5			
4			
4.5			
5			
5.5			
6			
6.5			
7			

4. The graph below shows the slope field for this differential equation from $x = 0$ through $x = 7$. Plot the y-values from Problems 1 through 3 on this graph paper. For which values of x does the numerical solution by Euler's method seem to follow the slope field? For which values of x is the numerical solution clearly wrong?

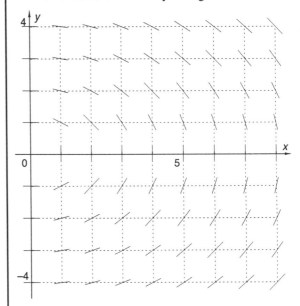

5. Solve the differential equation in Problem 1 algebraically. Plot the particular solution that contains (0, 3). Explain why Euler's method gives meaningless answers for larger values of x.

6. What did you learn as a result of doing this Exploration that you did not know before?

Calculus Explorations
© 1998 Key Curriculum Press

Exploration 46: A Predator-Prey Problem

<u>Objective</u>: Analyze the slope field for a differential equation modeling the populations of predatory coyotes and their deer prey.

The figure shows the slope field for a differential equation that represents the relative populations of deer, D, and coyotes, C, that prey on the deer. The slope at any point represents

$$\frac{dC}{dD} = \frac{dC/dt}{dD/dt}.$$

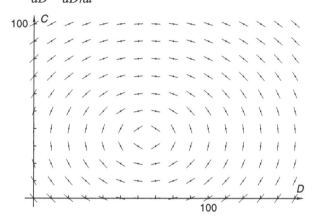

1. At the point (80, 20) there are relatively few coyotes. Would you expect the deer population to be increasing or decreasing at this point? Which way, then, would the graph of the solution start out under this condition?

2. Sketch the graph of the solution to the differential equation subject to the initial condition (80, 20). Describe what is expected to happen to the two populations under this condition.

3. Suppose that the initial number of deer is increased to 120, with the same 20 coyotes. Sketch the graph. What unfortunate result is predicted to happen under this condition?

4. Suppose that the initial deer population had been 140, with the same 20 coyotes. Sketch the graph. What other unfortunate thing would be expected to happen under this condition?

5. At the point (100, 35), what seems to be happening to the two populations? What seems to be happening at the point (65, 80)?

6. Does there seem to be an "equilibrium" condition for which neither population changes? Explain.

7. What did you learn as a result of doing this Exploration that you did not know before?

Exploration 47: Maxima, Minima, and Points of Inflection

<u>Objective</u>: Find the first and second derivatives for a function, and show how values of these correspond to the graph.

The figure shows the graph of

$$y = 5x^{2/3} - x^{5/3}$$

as it might appear on your grapher. In this Exploration you will make connections between the graph and the first and second derivatives.

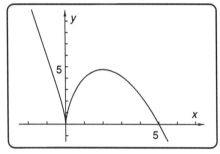

1. Find an equation for y'. Factor out any common factors, including $x^{-1/3}$ (the lowest power of x).

2. What value of x makes y' equal zero? What is true about the graph of y at that value of x?

3. You realize that zero raised to a negative power is infinite because it involves dividing by zero. What value of x makes y' infinite? What is true about the graph at this value of x?

4. Find an equation for the second derivative, y''. This is the derivative of the (first) derivative. Factor out any common factors.

5. If y'' is negative, the graph is **concave down.** Show algebraically that the graph is concave down at $x = 1$. Tell what it means graphically to be concave down.

6. Find a value of x at which y'' equals zero. The corresponding point on the graph is a **point of inflection.**

7. By picking a value of x on either side of the point of inflection in Problem 6, show that the graph is really concave up on one side and concave down on the other, even though the graph appears to be straight in a neighborhood of this x-value.

8. What did you learn as a result of doing this Exploration that you did not know before? (Over)

Calculus Explorations
© 1998 by Key Curriculum Press

Exploration 48: Derivatives and Integrals from Given Graphs

Objective: Given the graph of a function, sketch the graph of its derivative function or its integral function.

1. The graph below shows a function *h*. On the same axes sketch the graph of the derivative function, *h'*. At the cusp, $h'(1)$ is undefined. Make sure your graph accounts for this fact.

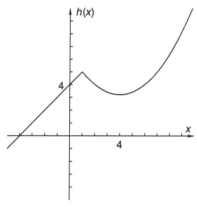

2. For *h* in Problem 1, is it also correct to say that $h'(1)$ is infinite? Explain.

3. For Problem 1, is *h continuous* at $x = 1$? Is *h differentiable* at $x = 1$?

4. The graph below shows the *derivative g'* of function *g*. On the same axes sketch the graph of *g*, using as an initial condition $g(0) = 3$.

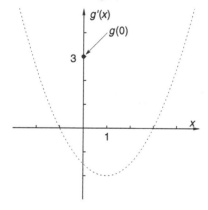

5. The graph below shows a function *y*. On the same axes sketch the graph of the derivative, *y'*.

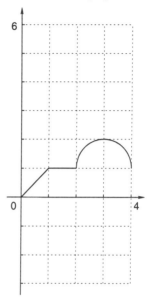

6. The graph shows the derivative of a function, *y'*. On the same axes sketch the graph of the function if (0, 0) is on the graph of *y*.

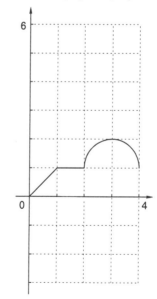

7. What did you learn as a result of doing this Exploration that you did not know before?

Exploration 49: Maximal Cylinder in a Cone Problem

<u>Objective</u>: Find an equation for a function, and maximize the function.

The figures show cylinders inscribed in a cone. The cone has radius 4 inches and altitude 12 inches. If the cylinder is tall and skinny as in the left-hand figure, the volume is small because the radius is small. If the cylinder is short and fat as in the right-hand figure, the volume is small because the altitude is small. In this Exploration you will learn how to find the dimensions that give the inscribed cylinder its *maximum* volume.

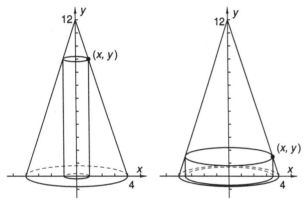

1. Calculate the volumes of cylinders for which the radius, x, equals 0, 1, 2, 3, and 4 units.

2. Write the volume of the cylinder in terms of the sample point (x, y) on an element of the cone. Then transform the equation so that the volume is in terms of x alone. You may find it useful to determine the equation of the line in Quadrant I that the sample point lies on.

3. Differentiate both sides of the volume equation in Problem 2 with respect to x.

4. As x increases the volume increases, reaches a maximum, then decreases again. At the maximum point its rate of change will be zero. Set dV/dx from Problem 3 equal to zero and solve for x.

5. Plot the graph of V. Sketch the graph here. Show on the graph that the maximum volume occurs at the value of x you found algebraically in Problem 4.

6. Find the radius, altitude, and volume of the maximum-volume cylinder.

7. Summarize the steps you went through in Problems 2 through 6 in order to find the maximum-volume cylinder.

8. What did you learn as a result of doing this Exploration that you did not know before? (Over)

Calculus Explorations
© 1998 Key Curriculum Press

Exploration 50: Volume by Plane Disk Slices

<u>Objective</u>: Calculate exactly the volume of a solid of variable cross-sectional area.

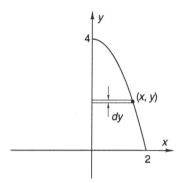

Draw the region.
Slice a strip perpendicular to the
axis of rotation.

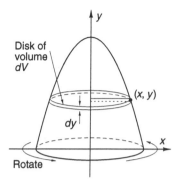

Rotate the region and the strip to
form a solid. The rotating strip forms
a flat disk.

The region in Quadrant I bounded by the parabola

$$y = 4 - x^2$$

is rotated about the y-axis to form a solid paraboloid, as shown above.

1. A representative slice of the region is shown in the left-hand figure. As the slice turns, it generates a **plane disk,** as shown in the right-hand figure. Let (x, y) be a sample point on the graph within the slice. Find the volume, dV, of the slice in terms of the coordinates of the sample point.

2. By appropriate algebra, transform dV so that it is terms of y alone.

3. Find, exactly, the volume V of the solid by adding the volumes of the slabs and taking the limit as the thickness of the slabs goes to zero (i.e., integrate).

4. As a rough check on your answer, compare it with the volume of a circumscribed cylinder and with the volume of an inscribed cone.

5. Based on your answer to Problem 4, make a conjecture about the volume of a paraboloid in relationship to the volume of the circumscribed cylinder.

6. What did you learn as a result of doing this Exploration that you did not know before?

Exploration 51: Volume by Plane Washer Slices

Objective: Find the volume of a solid of revolution that has a hole in it.

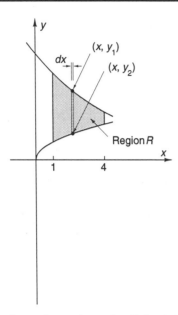

Region R

The figure above shows the region R that is bounded above and below by the graphs of

$$y_1 = 6e^{-0.2x} \quad \text{and} \quad y_2 = \sqrt{x},$$

and on the left and right by the lines $x = 1$ and $x = 4$. The figure below shows the solid generated by rotating R about the x-axis. A strip of width dx generates a **plane washer** of thickness dx whose cross-section is a **circular ring** of outer radius y_1 and inner radius y_2.

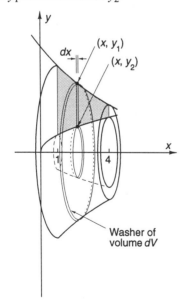

Washer of volume dV

1. Write the area of the circular ring as a function of y_1 and y_2. Use the result to write the volume, dV, of the washer.

2. Transform the equation for dV so that it is in terms of x alone.

3. Find V by adding up the dV values and taking the limit (i.e., integrate). Use the fundamental theorem to evaluate the integral *exactly*.

4. Get a decimal approximation for the volume you found in Problem 3.

5. Do the integration in Problem 3 numerically. Show that the answer is (virtually) the same as in Problem 4.

6. What did you learn as a result of doing this Exploration that you did not know before? (Over)

Calculus Explorations
© 1998 Key Curriculum Press

Exploration 52: Volume by Cylindrical Shells

<u>Objective</u>: Find the volume of a solid of revolution by slicing the rotated region *parallel* to the axis of rotation.

The left-hand figure below shows the region under the graph of $y = 4x - x^2$ from $x = 0$ to $x = 3$. The right-hand figure shows the solid formed when this region is rotated about the y-axis.

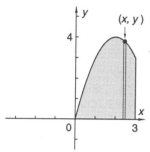 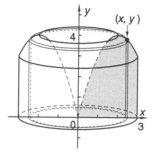

1. As the region rotates, the strip parallel to the y-axis generates a **cylindrical shell** as shown below. The volume, dV, of this shell equals the circumference at the sample point (x, y), times the altitude of the shell, times the thickness of the shell, dx. Write an equation for dV in terms of x, y, and dx.

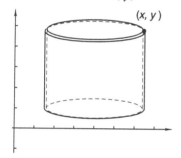

2. Substitute for y to get dV in terms of x and dx alone.

3. Calculate the volume of the solid by adding up all the dV's and taking the limit (i.e., integrate). Tell why the limits of integration are 0 to 3, not –3 to 3.

4. Demonstrate that you understand slicing a solid into cylindrical shells by using the technique to find the volume of the solid shown below.

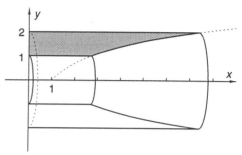

In this solid the region bounded by the y-axis; the lines $y = 1$, and $y = 2$; and the graph of $y = \ln x$ is rotated about the x-axis to form a solid. (For clarity, only the back side of the solid is shown.)

5. What did you learn as a result of doing this Exploration that you did not know before?

Exploration 53: Length of a Plane Curve (Arc Length)

<u>Objective</u>: Calculate the arc length of a plane curve approximately by geometry, and exactly by calculus.

The figure shows the graph of $y = x^2$ from $x = -1$ to $x = 2$. In this Exploration you will find the **arc length** of this graph segment.

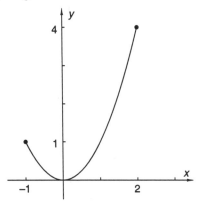

1. Draw line segments connecting (consecutively) the points on the graph where x is -1, 0, 1, and 2. Estimate the length of the graph by calculating the lengths of these segments. Does this overestimate or underestimate the arc length?

2. Find an estimate of the length using segments connecting the points on the graph where x is -1, -0.5, 0, 0.5, 1, 1.5, and 2. Explain why this estimate is better than the one in Problem 1.

3. In Problems 1 and 2 the arc length, L, is approximated by the sum

$$L \approx \Sigma \sqrt{\Delta x^2 + \Delta y^2} \,.$$

In Problem 1, $\Delta x = 1$ and in Problem 2, $\Delta x = 0.5$. What would one have to do to find the *exact* value of L?

4. If $y = f(x)$, and f is a differentiable function, then any term in the sum above can be written exactly as

$$\Delta L = \sqrt{1 + [f'(c)]^2} \; \Delta x$$

where c is some number in the respective subinterval. Consult the text, then tell what theorem is the basis for this fact.

5. Explain why, in the limit, ΔL can be written as

$$dL = \sqrt{dx^2 + dy^2} \,.$$

6. Find dL for $y = x^2$. Then write L exactly as a definite integral.

7. Find a decimal approximation for the exact arc length by evaluating the integral in Problem 6 numerically.

8. What did you learn as a result of doing this Exploration that you did not know before? (Over)

Calculus Explorations
© 1998 Key Curriculum Press

Exploration 54: Area of a Surface of Revolution

<u>Objective</u>: Calculate exactly the area of a doubly-curved surface of revolution.

If a line segment is rotated about an axis coplanar with it, a *singly* curved surface is formed (a cone or cylinder). Such a surface can be rolled out flat and its area found by calculus or geometry techniques you have learned before. If a plane curve is rotated about an axis coplanar with it, a *doubly* curved surface is formed. Spheres, ellipsoids, paraboloids, etc., are examples. Since such a surface *cannot* be flattened without distortion, its area must be found without first flattening. The figure below shows a doubly-curved surface formed by rotating a graph about the *x*-axis.

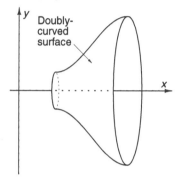

Rotate a curved graph,
get a doubly-curved surface.

In the problems below you will turn the new problem of finding the area of a doubly-curved surface into the old problem of finding areas of many singly-curved surfaces. Then you will add up the areas of these surfaces and take the limit (i.e., integrate).

1. Suppose that a curve in the figure above has been sliced into curved segments by partitioning the *x*-axis into subintervals of equal length. Show on the diagram how rotating one of these segments gives a surface that resembles a frustum of a cone.

2. Prove that the lateral area of a cone is $\pi R L$, where R is the radius of the cone's base and L is the slant height of the cone.

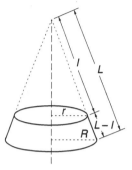

3. The diagram above shows a frustum of a cone with base radii and slant heights R and r, and L and l. Prove that the lateral area of the frustum equals the circumference at the *average* radius multiplied by the slant height of the frustum.

4. The slice you drew in Problem 1 has a slant height approximately equal to dL, the differential of arc length. Write an equation for dS, the differential of surface area.

5. The graph of $y = x^3$ from $x = 0$ to $x = 1$ is rotated about the *x*-axis to form a surface. Find its area.

6. What did you learn as a result of doing this Exploration that you did not know before? (Over)

Exploration 55: Area of an Ellipse in Polar Coordinates

<u>Objective</u>: Find the area of an ellipse from its polar equation, then compare with the area found by familiar geometry.

The figure shows the **ellipse** in polar coordinates

$$r = \frac{10}{3 - 2\cos\theta}.$$

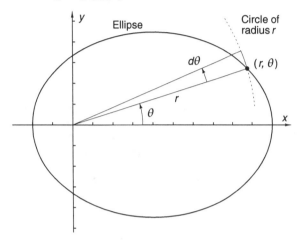

1. Set your grapher to POLAR mode and plot the graph. Use equal scales on both x- and y-axes. Does your graph agree with the figure above?

2. The sample point (r, θ) shown in the figure is at $\theta = 0.3$ radian. Calculate r, x, and y for this point. Show that all three agree with the graph.

3. The area of a wedge-shaped piece of the elliptical region bounded by the graph is approximately equal to the area of the sector of a circle. Calculate the area of the sector shown in the figure above, if $\theta = 0.3$ radian and $d\theta = 0.1$ radian.

4. Show that in general, the area dA of the sector is

$$dA = \frac{1}{2}r^2\,d\theta.$$

5. The exact area of the elliptical region is the *limit* of the *sum* of the sectors' areas. That is, the area equals the definite integral of dA. Write an integral representing the exact area. Evaluate the integral numerically.

6. The x-radius, a, of the ellipse shown is 6 units. Measure or calculate the y-radius, b. Then confirm that the answer you got by integration in Problem 5 agrees with the answer you get using the ellipse area formula $A = \pi ab$.

7. What did you learn as a result of doing this Exploration that you did not know before? (Over)

Exploration 56: Integration by Parts Practice

Objective: Integrate by parts to find antiderivatives quickly.

Find the integral. Show the steps you take.

1. $\int x^2 \sin 3x \, dx$

2. $\int x^5 \ln 4x \, dx$

3. $\int e^{5x} \cos 6x \, dx$

4. $\int x \, (\ln x)^3 \, dx$

5. $\int \sin^{10} x \cos x \, dx$ (Be clever!)

6. $\int \sin^{10} x \, dx$, in terms of $\int \sin^8 x \, dx$

7. What did you learn as a result of doing this Exploration that you did not know before? (Over)

Exploration 57: Reduction Formulas

<u>Objective</u>: Integrate by parts to find a reduction formula, and use that formula in a volume problem.

1. Integrate by parts to find a formula for $\int \sin^{10} x\, dx$ in terms of $\int \sin^8 x\, dx$.

4. Problem 3 involves $\int \sin^6 x\, dx$. Use the reduction formula to evaluate this indefinite integral. Use the result to find the volume of the solid algebraically, using the fundamental theorem.

2. Based on the pattern you observe in the answer to Problem 1, write a formula for $\int \sin^n x\, dx$ in terms of $\int \sin^{n-2} x\, dx$.

3. The graph shows $y = \sin^3 x$. Find numerically the volume of the solid generated by rotating around the x-axis the region under this graph from $x = 0$ to $x = \pi$.

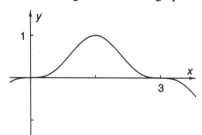

5. What did you learn as a result of doing this Exploration that you did not know before?

Calculus Explorations
© 1998 Key Curriculum Press

Exploration 58: Integrals of Special Powers of Trigonometric Functions

Objective: Integrate odd powers of sine and cosine, and even power of secant and cosecant.

1. Integrate $\int \cos^7 x \sin x \, dx$ as a power.

2. Explain why $\int \cos^7 x \, dx$ cannot be integrated as a power.

3. Associate all but one of the cosines in the integral in Problem 2, change them to sines using the Pythagorean properties, then evaluate the integral as a *sum* of *powers*.

4. Explain why $\int \cos^6 x \, dx$ cannot be integrated by the method of Problem 3.

5. Integrate $\int \tan^5 x \sec^2 x \, dx$ as a power.

6. Explain why $\int \sec^8 x \, dx$ cannot be integrated as a power.

7. Associate all but two of the secants in the integral in Problem 6, change them to tangents using the Pythagorean properties, then evaluate the integral as a *sum* of *powers*.

8. Explain why $\int \sec^7 x \, dx$ cannot be integrated by the method of Problem 7.

9. What did you learn as a result of doing this Exploration that you did not know before? (Over)

Exploration 59: Other Special Trigonometric Integrals

<u>Objective</u>: Find the antiderivatives $\int \sin^2 x\, dx$ and $\int \cos^2 x\, dx$ and $\int \cos ax \sin bx\, dx$.

1. You recall that

$$\cos (A + B) = \cos A \cos B - \sin A \sin B.$$

   Use this property to write the double argument property for $\cos 2x$.

2. Transform the double argument property so that $\cos 2x$ is expressed in terms of $\cos x$ alone.

3. Transform the double argument property so that $\cos 2x$ is expressed in terms of $\sin x$ alone.

4. Transform the properties in Problems 2 and 3 so that $\cos^2 x$ is expressed in terms of $\cos 2x$, and so that $\sin^2 x$ is expressed in terms of $\cos 2x$.

5. Use the results of Problem 4 to find the integrals $\int \cos^2 x\, dx$ and $\int \sin^2 x\, dx$.

6. Products of sine and cosine with two different arguments can be integrated by parts. Integrate:

   $\int \cos 7x \sin 5x\, dx$

7. Use the result of Problem 6 to write a formula for $\int \cos ax \sin bx\, dx$.

8. What did you learn as a result of doing this Exploration that you did not know before?

Calculus Explorations
© 1998 Key Curriculum Press

Exploration 60: Introduction to Integration by Trigonometric Substitution

<u>Objective</u>: Integrate a square root of a quadratic by making a rationalizing substitution.

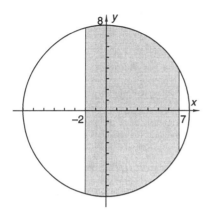

1. The diagram above shows the circle for which

$$y = \pm \sqrt{64 - x^2}.$$

Use the most time-efficient method to find the area of the zone of this circle from $x = -2$ to $x = 7$.

2. In Problem 1 you evaluated

$$\int \sqrt{64 - x^2} \, dx$$

numerically. Although the square root can be written as the 1/2 power, the integral cannot be done algebraically as a power. Why not?

3. The radical in Problems 1 and 2 looks like the third side of a right triangle (diagram below) with one leg x, hypotenuse 8, and θ in standard position.

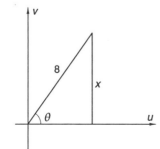

Write the integral in Problem 2 in terms of trig functions of θ.

4. Using the techniques you have learned for integrating powers of trig functions, evaluate the integral in Problem 3 algebraically.

5. Do the reverse substitution to get the answer in Problem 4 into terms of x.

6. Use the answer to Problem 5 to find *exactly* the area of the zone in Problem 1. Show that a decimal approximation of this exact answer agrees with the answer found numerically.

7. What did you learn as a result of doing this Exploration that you did not know before? (Over)

Exploration 61: Integrals of Rational Functions by Partial Fractions

<u>Objective</u>: Integrate algebraically a function of the form (polynomial)/(polynomial).

1. The integral

$$\int \frac{10x - 32}{x^2 - 4x - 5}\,dx$$

has an integrand that is a **rational algebraic function.** It can be transformed to

$$\int \frac{10x - 32}{(x - 5)(x + 1)}\,dx$$

The integrand can then be broken up into two **partial fractions,**

$$\frac{A}{x - 5} + \frac{B}{x + 1}.$$

Find the constants A and B.

2. The integral in Problem 1 can be done as a sum of two terms, each of which has the form of the **reciprocal function.** Find the indefinite integral.

3. Ask your instructor to show you the way to find the partial fractions in *one* step, in your head.

4. Study Heaviside's method to see why the shortcut works. Check here when you feel you understand it well enough to explain it to other members of your study group. _____

5. Heaviside's method and its shortcut can be extended to integrals with any number of distinct linear factors in the denominator, as long as the numerator is of degree at least one lower than the denominator. Do this integration:

$$\int \frac{11x^2 - 22x - 13}{x^3 - 2x^2 - 5x + 6}\,dx$$

6. What did you learn as a result of doing this Exploration that you did not know before? (Over)

Calculus Explorations
© 1998 Key Curriculum Press

Exploration 62: Chain Experiment

<u>Objective</u>: Find the particular equation for a hanging chain, and verify it by actual measurement.

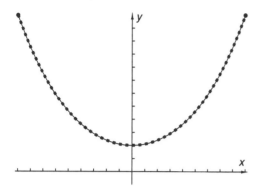

1. Hang a chain from the frame around the chalkboard, as shown in the diagram. Let x be the horizontal distance (cm) from the vertex to a point on the chain, and let y be the vertical distance (cm) from the chalk tray to the point. Measure x and y for the vertex and for the two endpoints.

    Vertex: _____

    Ends: Left: _____ Right: _____

2. The general equation for a hanging chain is

    $$y = \frac{h}{w} \cosh \frac{w}{h} x + C,$$

    where h is the horizontal tension in the chain and w is the weight of the chain per unit length. Let $k = h/w$. Calculate the constants k and C. Show your work. Store the answers in your grapher.

3. Weigh and measure the chain. Use the results to find values of h and w.

4. Calculate y for each 20 cm from the vertex, from the left end to the right end of the chain. Use a time-efficient method. Round to one decimal place.

x	y		x	y

5. Remove the chain from the board. Plot the data on the board as accurately as you can. Then hang the chain again. How closely does the calculated data fit the shape of the actual chain?

6. Use the equation to calculate the length of the chain between the two endpoints. Then measure the chain to see how close your calculated value is to the actual value.

7. On the back of this sheet, derive the equation

    $$y = \frac{h}{w} \cosh \frac{w}{h} x + C,$$

    using the fact that the tension in the chain is equal to the vector sum of the horizontal tension (of magnitude h) and the vertical tension, and is always directed along the chain.

8. What did you learn as a result of doing this Exploration that you did not know before?

Exploration 63: Introduction to Improper Integrals

<u>Objective</u>: Find the limit of a definite integral as the upper limit of integration becomes infinite.

Calvin is driving at 80 ft/sec (about 55 mph). At time $t = 0$ seconds he lifts his foot from the accelerator and coasts. His velocity decreases as shown in the graph.

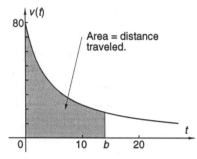

1. Calvin figures that his velocity is given by

   $$v(t) = 320(t + 4)^{-1}.$$

   How far has he gone after 10, 20, and 50 seconds?

2. Find algebraically the distance Calvin has gone after $t = b$ seconds. Use the result to find out how long it will take him to go a distance of 1000 ft.

Phoebe does the same thing Calvin did. Her velocity-time graph is shown below.

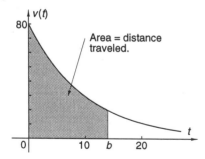

3. Phoebe figures that her velocity is given by

   $$v(t) = 80e^{-0.1t}.$$

   How far has she gone after 10, 20, and 50 seconds?

4. Find algebraically the distance Phoebe has gone after $t = b$ seconds. Use the result to find out how long it will take her to go a distance of 1000 ft.

5. Show that Phoebe's distance approaches a limit as b approaches infinity, but Calvin's distance does *not*.

6. How do you reconcile the fact that Phoebe's distance approaches a limit, and the fact that her velocity never reaches zero?

7. What did you learn as a result of doing this Exploration that you did not know before? (Over)

Calculus Explorations
© 1998 Key Curriculum Press

Exploration 64: Miscellaneous Integration Practice!

<u>Objective</u>: Evaluate an antiderivative when the particular technique is not specified.

1. $\int \tan^5 4x \, dx$

4. $\int_1^\infty x^3 \, e^{-x} \, dx$

2. $\int \sqrt{1 + t^2} \, dt$

5. $\int \dfrac{x}{(x - 2)(x - 3)(x - 4)} \, dx$

3. $\int \tanh x \, dx$

(Over)

Exploration 64, continued

6. $\int \sin^5 x \, dx$

7. $\int (x^4 + 2)^3 \, dx$

8. $\int x^2 \, e^{x^3} \, dx$

9. $\int e^{ax} \cos bx \, dx$

10. $\int \sin^{-1} ax \, dx$

11. What did you learn as a result of doing this Exploration that you did not know before?

Exploration 65: Finding Distance from Acceleration Data

Objective: Given data for the acceleration of a moving object, find its displacement at various times.

The following table shows acceleration of a moving object, in (m/sec)/sec.

t	a		
0	5		
10	12		
20	11		
30	−4		
40	−13		
50	−20		
60	0		

1. Assume that the acceleration changes linearly in each 10-second time interval, and that the velocity at time $t = 0$ is 3 m/sec. Show how the velocity may be calculated at time $t = 10$.

2. In the table above, put in a column showing the velocity at the end of each 10-second time interval.

3. Assume that the average velocity in each time interval is the average of the velocity at the beginning and the velocity at the end of the interval. Assume also that the displacement is zero when $t = 0$. Show how the displacement may be calculated at time $t = 10$ seconds.

4. In the table, put in a column that shows the displacement at the end of each 10-second time interval.

5. Explain how you can tell that sometime between $t = 0$ and $t = 60$ the object stops and begins going backwards.

6. Does the object go far enough backwards to go back beyond its starting point? Explain.

7. What did you learn as a result of doing this Exploration that you did not know before?

Exploration 66: Average Velocity

Objective: Given an equation for the velocity of a moving object, calculate the average velocity over a given time interval.

Rhoda Honda accelerates her motorcycle from one stop sign, then slows down at the next stop sign. She figures that her velocity is given by

$$v(t) = 6t - 0.3t^2,$$

where t is in seconds and $v(t)$ is in feet per second. The velocity-time graph is shown below.

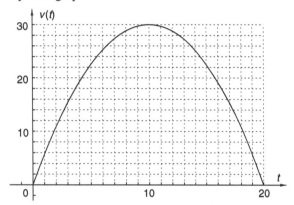

1. Find Rhoda's velocity at $t = 1$ and $t = 17$. Show that the graph agrees with these numbers.

2. Find Rhoda's displacement from $t = 1$ to $t = 17$.

3. Rhoda's **average velocity** in the interval [1, 17] is her displacement for that interval divided by the number of seconds in that interval. Find her average velocity.

4. Does the average velocity over the interval [1, 17] equal the average of $v(1)$ and $v(17)$?

5. Shade lightly the region under the graph that represents Rhoda's displacement.

6. Draw a horizontal line across the graph at the average velocity you calculated in Problem 3. Then draw a rectangle bounded by this line, the t-axis, and the lines $t = 1$ and $t = 17$. How does the area of this rectangle compare with the area of the region you shaded in Problem 5?

7. Make another conclusion about the areas of various regions in the figure that results from Problem 6.

8. Show that you understand the conclusions you have made above by estimating graphically the average velocity for this function over [0, 10].

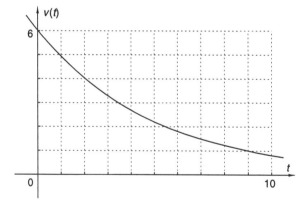

9. What did you learn as a result of doing this Exploration that you did not know before? (Over)

Exploration 67: Introduction to Related Rates

<u>Objective</u>: Given the rates of change for two variable quantities, find the rate of change for another related quantity.

Phoebe is going east on Alamo Street at 40 ft/sec. Meanwhile, Calvin is going south on Heights Street at 30 ft/sec (see figure). Phoebe wishes to determine whether the straight-line distance between them is increasing or decreasing.

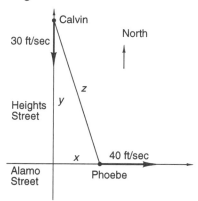

1. Let x be Phoebe's distance from the intersection and let y be Calvin's distance. Write the rates you **know.** Explain why dy/dt is negative.

   <u>Know</u>: $\dfrac{dx}{dt} =$ _____, $\dfrac{dy}{dt} =$ _____.

2. Let z be the straight-line distance between Calvin and Phoebe. Write the symbol for the rate you **want.**

   <u>Want</u>: _____

3. To find the rate you want, you need an equation relating the variables in the known and wanted rates. Write an equation relating x, y, and z.

4. Differentiate both sides of the equation in Problem 3 implicitly with respect to t. Solve the resulting equation for dz/dt, the wanted rate.

5. When Phoebe is 200 feet from the intersection, Calvin is 600 feet from the intersection. At this time is the straight-line distance between them increasing or decreasing? At what rate?

6. Naive thinking might lead you to conclude that since Phoebe is going faster *away* from the intersection than Calvin is going *toward* the intersection, the distance between them is *increasing*. Do the calculations confirm or refute this conclusion?

7. Write a list of the things you did in solving this **related rates** problem.

8. What did you learn as a result of doing this Exploration that you did not know before?

Exploration 68: Introduction to Minimal Path Problems

<u>Objective</u>: Find out how to construct a road of minimal cost between points in different regions.

Anita Hammer builds a cabin in the woods. She wants to make a road to the cabin starting at the gate, going through the clearing along the edge of the woods, and then cutting through the woods to the cabin (see figure).

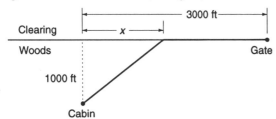

1. The distance from the cabin to the point on the edge of the clearing closest to the cabin is 1000 feet. The distance from the gate to this closest point is 3000 feet. Building the roadway costs $20 per foot through the clearing and $30 per foot through the woods. Let *x* be the number of feet from the closest point to the point where the road will cut off. Write an equation expressing cost, *C*, of the road in terms of *x*.

2. Plot the graph of cost versus *x*. Sketch the result here.

3. At what value of *x* does the cost seem to be a minimum?

4. Differentiate to find dC/dx algebraically. Set the derivative equal to zero and solve for *x*. Why does the answer tell you where the minimum value of *C* is?

5. What is the minimum cost of the road? How much more would the road cost if Anita built it from the cabin to the closest point on the edge of the clearing and thence to the gate? How much more would it cost if she built it directly from the gate to the cabin?

6. What did you learn as a result of doing this Exploration that you did not know before?

Calculus Explorations
© 1998 Key Curriculum Press

Exploration 69: Introduction to Calculus of Vectors

<u>Objective</u>: Find the derivative function for an object moving in two dimensions.

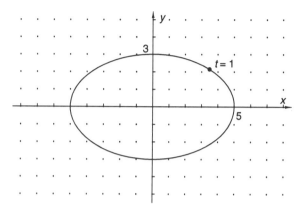

1. The graph above shows the path of a pendulum moving in both the x- and y-directions. The position (in feet) at any time t seconds is given by the parametric equations

   $x = 5 \cos 0.8t$

   $y = 3 \sin 0.8t.$

   Find dx/dt and dy/dt when $t = 1$.

2. The derivatives in Problem 1 are the velocity of the pendulum in the x- and y-directions. On the figure, construct vectors of lengths dx/dt and dy/dt in the x- and y-directions, respectively, starting at the point on the graph where $t = 1$.

3. On the figure, construct the vector sum of the two vectors in Problem 1. What relationship does this vector have to the graph?

4. The vector sum in Problem 3 is the velocity vector of the pendulum. How fast is the pendulum moving at time $t = 1$?

5. Find the second derivatives of x and y with respect to t at time $t = 1$. On the figure, construct vectors of these lengths in the x- and y-directions, respectively, starting at the point on the graph where $t = 1$. Construct the resultant of these vectors.

6. The resultant vector in Problem 5 is the acceleration of the pendulum. Based on your work, is the pendulum speeding up or slowing down at $t = 1$? How can you tell?

7. Just for fun, see if you can figure out the *rate* at which the pendulum is speeding up or slowing down when $t = 1$.

8. What did you learn as a result of doing this Exploration that you did not know before?

Exploration 70: Derivatives of a Position Vector

Objective: Given the vector equation for motion in two dimensions, find velocity and acceleration as functions of time.

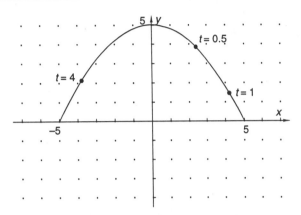

The graph above shows the path traced by a moving object with position vector $\vec{r}(t)$ given by

$$\vec{r}(t) = (5 \sin t)\,\vec{i} + (5 \cos^2 t)\,\vec{j}.$$

Distances x and y are in feet and time t is in seconds.

1. Confirm that the points shown at $t = 0.5$ and $t = 1$ are correct.

2. On the graph above, construct the position vectors $\vec{r}(0.5)$ and $\vec{r}(1)$.

3. Calculate the difference $\Delta\vec{r} = (\vec{r}(1) - \vec{r}(0.5))$ by subtracting the respective components. Construct $\Delta\vec{r}$ with its tail at the head of $\vec{r}(0.5)$. Where is the head of $\Delta\vec{r}$?

4. The **average velocity** vector of the object for interval $[0.5, t]$ is the difference quotient

$$\vec{v}_{av} = \frac{\vec{r}(t) - \vec{r}(0.5)}{t - 0.5}.$$

Find the average velocities for $[0.5, 1]$ and $[0.5, 0.6]$. Construct these vectors starting at the head of the position vector $\vec{r}(0.5)$.

5. Find the (instantaneous) velocity vector, $\vec{v}(0.5)$. Construct it starting at the end of $\vec{r}(0.5)$. How does $\vec{v}(0.5)$ relate to the graph? How do the average velocity vectors in Problem 4 relate to $\vec{v}(0.5)$?

6. Find the speed of the object at time $t = 0.5$.

7. Write vector equations for the velocity vector, $\vec{v}(t)$, and the acceleration vector, $\vec{a}(t)$.

(Over)

Exploration 70, continued

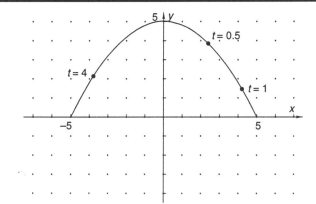

8. On this copy of the graph, construct the position vector $\vec{r}(4)$. Calculate the vectors $\vec{v}(4)$ and $\vec{a}(4)$. From the head of $\vec{r}(4)$, construct the velocity and acceleration vectors.

9. How fast is the object going at time $t = 4$?

10. Compute $\vec{a}_t(4)$, the **tangential component** of the acceleration (parallel to the path). Construct this vector starting at the position of the object at time $t = 4$, thus showing that it really is tangent.

11. Is the object speeding up or slowing down at $t = 4$? How do you tell? At what rate is it speeding up or slowing down?

12. Compute $\vec{a}_n(4)$, the **normal component** of acceleration (perpendicular to the path). Construct this vector starting at the position of the object at time $t = 4$, thus showing that it really is perpendicular to the path.

13. Toward which side of the path does $\vec{a}_n(4)$ point? What effect does this component have on the motion of the object?

14. What did you learn as a result of doing this Exploration that you did not know before?

Exploration 71: Hypocycloid Vector Problem

<u>Objective</u>: Given the vector equation for motion in two dimensions, find velocity and acceleration as functions of time.

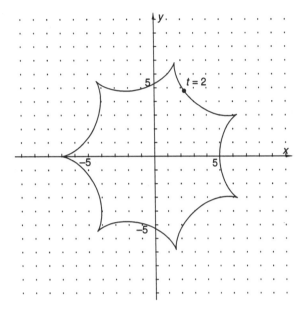

5. Geometrically, does the object seem to be speeding up or slowing down when $t = 2$? How do you tell?

6. Find the scalar projection of $\vec{a}(2)$ on $\vec{v}(2)$.

7. Algebraically, is the object speeding up or slowing down when $t = 2$? How do you tell?

The graph above shows a **hypocycloid of seven cusps**. The graph is the path traced by a point on a circle that rolls on the inside of another circle that has 7 times the radius. The position vector $\vec{r}(t)$ is given by

$$\vec{r}(t) = (6 \cos 0.5t - \cos 3t)\,\vec{i} + (6 \sin 0.5t + \sin 3t)\,\vec{j}$$

Distances x and y are in feet and time t is in seconds,

1. Plot the path of the object on your grapher. Does the graph agree with the one shown here?

2. Write equations for $\vec{v}(t)$ and for $\vec{a}(t)$, the velocity and acceleration vectors.

8. Calculate the tangential component of acceleration at $t = 2$.

9. Calculate the normal component of acceleration at $t = 2$.

3. Find $\vec{v}(2)$ and $\vec{a}(2)$. Round to two places.

10. Construct the tangential and normal components of acceleration on the graph shown.

4. Construct $\vec{v}(2)$ and $\vec{a}(2)$ on the graph above, starting at the head of $\vec{r}(2)$.

(Over)

Calculus Explorations
© 1998 Key Curriculum Press

Exploration 71, continued

11. Tell the physical effects of the tangential and normal components of acceleration at $t = 2$.

12. There is a cusp at the point with approximate coordinates $(6, 3)$. What must be true of dx/dt and dy/dt at this point?

13. Find by appropriate numerical techniques the value of t at the cusp near $(6, 3)$. Leave a "trail" so that the reader can tell how you did this.

14. Find a point close to $(6, 3)$ where $dy/dt = 0$ but $dx/dt \neq 0$. What is the value of t at that point?

15. How fast is the object moving when $t = 2$?

16. How far does the object travel in one complete revolution? How does this number compare with a circle of an appropriate radius?

17. What did you learn as a result of doing this Exploration that you did not know before?

Exploration 72: Pumping Work Problem

<u>Objective</u>: Find the work done in pumping a liquid from a container to a point above the container.

Treadwell Winery has a vat on the ground floor where they store wine. They pump the wine up to the second floor, 15 feet above the bottom of the vat, where it is put into bottles. The vat is the shape of a truncated cone with base radius 6 feet and top radius 8 feet. The vat is 10 feet deep. Mr. Treadwell calls upon you to find out how much work is done in pumping a full vat of wine up to the second floor.

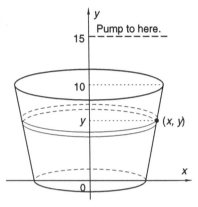

1. The work done by a pump in lifting a horizontal "slice" of the wine from y feet up to 15 feet equals the force needed to lift the slice multiplied by the distance the slice is lifted. The force equals the weight of the slice, which is equal to its volume times its density. Write dW, the work needed to lift the slice, in terms of the sample point (x, y) shown in the figure. Assume that the wine weighs 63 pounds per cubic foot, about the same as water.

2. To get a relationship between x and y, observe that (x, y) is a point on a line in the xy-plane. Use this information to write dW in terms of y alone.

3. By appropriate calculus, find the work done in pumping all of the wine from the vat up to the second floor.

4. What did you learn as a result of doing this Exploration that you did not know before?

Exploration 73: Mass of a Variable-Density Solid

<u>Objective</u>: Find the mass of a solid of revolution if its density varies axially or radially.

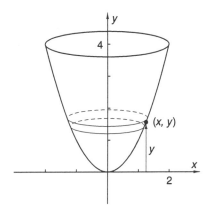

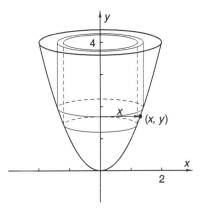

1. The figure shows the paraboloid formed by rotating about the y-axis the region above the graph of $y = x^2$, below $y = 4$, and to the right of the y-axis, where x and y are in centimeters. Assume that the density of the solid varies **axially** (in the direction of the axis of the solid), being equal to $3y^{1/2}$ grams per cubic centimeter at a sample point (x, y) in a horizontal disk. Find the mass, dm, of the disk in terms of the sample point. Then use appropriate calculus to find the mass of the entire solid.

2. The figure shows another solid congruent to the solid in Problem 1. The density of this solid varies **radially** (in the direction of the radius), being equal to $x + 5$ grams per cubic centimeter at a sample point (x, y) in a cylindrical shell. Find the mass, dm, of the cylindrical shell in terms of the sample point. Then use appropriate calculus to find the mass of the entire solid.

3. What did you learn as a result of doing this Exploration that you did not know before?

Exploration 74: Moment of Volume, and Centroid

<u>Objective</u>: Find the first moment of volume of a solid, and find the solid's geometrical center (centroid).

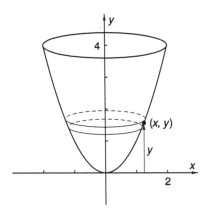

1. The figure shows the paraboloid formed by rotating about the y-axis the region above the graph of $y = x^2$, below $y = 4$, and to the right of the y-axis, where x and y are in centimeters. Find the volume, V, of the solid. (You may do this quickly!)

2. The **first moment of volume** of the solid with respect to the xz-plane is defined to be the volume of the solid times the displacement of the solid from the xz-plane. (The z-axis comes straight out of the page.) Unfortunately, different parts of the solid are at different displacements from the plane. Explain why the (first) moment of the horizontal disk shown in the figure can be found approximately by multiplying the volume of the disk by the y-coordinate of the sample point.

3. Find the moment, dM_{xz}, of the disk in terms of the sample point (x, y). Then do appropriate algebra and calculus to find the moment, M_{xz}, of the entire solid with respect to the xz-plane.

4. The **centroid** (which means "centerlike") of the solid is at the point $\bar{y}$ (pronounced "y bar") on the y-axis for which

$$M_{xz} = \bar{y} \cdot V.$$

Find $\bar{y}$. Mark this point on the figure given. Explain why it is *more* than halfway up the solid.

5. What did you learn as a result of doing this Exploration that you did not know before?

Calculus Explorations
© 1998 Key Curriculum Press

Exploration 75: Moment of Mass and Center of Mass

<u>Objective</u>: Learn the meaning of moment of mass and center of mass.

The figure shows the paraboloid formed by rotating about the y-axis the region above the graph of $y = x^2$, below $y = 4$, and to the right of the y-axis, where x and y are in centimeters. Assume that the density of the solid varies **axially** (in the direction of the axis of the solid), being equal to $3y^{1/2}$ grams per cubic centimeter at a sample point (x, y) in a horizontal disk.

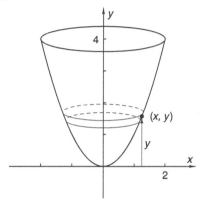

1. Find the mass of the solid. If you have worked Exploration 73, you may use the results of that exploration.

2. The **first moment of mass** of the solid with respect to the xz-plane is defined to be the mass of the solid times the displacement of the solid from the xz-plane. (The z-axis comes straight out of the page.) Unfortunately, different parts of the solid are at different displacements from the plane. Explain why the (first) moment of mass of the horizontal disk shown in the figure can be found approximately by multiplying the mass of the disk by the y-coordinate of the sample point.

3. Find the moment of mass, dM_{xz}, of the disk in terms of the sample point (x, y). Then do appropriate algebra and calculus to find the moment of mass, M_{xz}, of the entire solid with respect to the xz-plane.

4. The **center of mass** of the solid is the point at which the mass could be concentrated to produce the same moment of mass with respect to any plane. It is at the point $\bar{y}$ on the y-axis for which

$$M_{xz} = \bar{y} \cdot m.$$

Find $\bar{y}$. Mark this point on the figure given.

5. In Exploration 74 you may have found that the **centroid** (center of volume) is at $y = 8/3$, which is *not* the same point as the center of mass. Under what condition would the center of mass be at the centroid?

6. What did you learn as a result of doing this Exploration that you did not know before? (Over)

Exploration 76: Second Moment of Mass

<u>Objective</u>: Learn the meaning of second moment of mass and of radius of gyration.

The figure below shows an aluminum solid. It is in the shape of the paraboloid formed by rotating about the y-axis the region above the graph of $y = x^2$, below $y = 4$, and to the right of the y-axis, where x and y are in centimeters. Its density is 2.7 g/cm³.

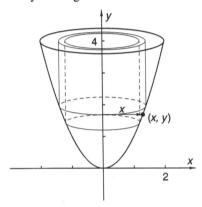

1. Find the mass, m, of the solid. (You may do this quickly!)

2. The **second moment of mass** of the solid with respect to the y-axis is defined to be the mass of the solid times the *square* of its distance from the y-axis. This quantity is sometimes called the **moment of inertia.** It measures how difficult it is to change the angular velocity of the solid as it rotates about the y-axis. Explain why it is appropriate to slice the solid into cylindrical shells in order to find dM_{2y}, the differential of second moment.

3. Find the second moment of mass, dM_{2y}, of a cylindrical shell in terms of the sample point (x, y). Then do appropriate algebra and calculus to find the second moment of mass, M_{2y}, of the entire solid with respect to the y-axis.

4. The **radius of gyration** of the solid is the distance from the y-axis at which the mass could be concentrated to produce the same second moment. That is, it is the distance $\bar{r}$ for which

$$M_{2y} = \bar{r}^2 m.$$

Find the radius of gyration.

5. Suppose that a weight on a string is being rotated around. If the weight is moved three times as far from the axis of rotation as it had been, by what factor is the second moment of mass changed?

6. What did you learn as a result of doing this Exploration that you did not know before? (Over)

Calculus Explorations
© 1998 Key Curriculum Press

Exploration 77: Force Exerted by a Variable Pressure

<u>Objective</u>: Find the force exerted on a surface if the pressure is different at various places on the surface.

<u>Dam Problem</u>: A dam is to be built across a gully. The back face of the dam (where the water will be) is the shape of the parabolic region bounded below by the graph of $y = 0.03x^2$ and above by the line $y = 12$, where x and y are in feet (see figure). The **force** the water will exert on the dam face equals the water pressure (pounds per square foot) times the area of the dam face. Unfortunately, the pressure gets higher as the water gets deeper, so the force cannot be found by simple multiplication.

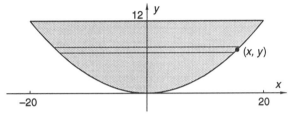

1. The pressure, p, equals the depth of the water times the weight density of water, 62.4 pounds per cubic foot. Find the pressure at the sample point (x, y) shown in the figure when the water is all the way to the top of the dam, $y = 12$ feet.

2. The pressure acting at any point in a horizontal strip such as the one shown in the figure is approximately equal to the pressure at the sample point. Find the force, dF, acting on the strip in terms of the sample point (x, y).

3. By appropriate algebra and calculus, find the total force, F, acting on the back face of the dam.

4. The **first moment of force** with respect to the x-axis is defined to be the force times the displacement from the x-axis. Find the first moment of force, M_x, with respect to the x-axis.

5. The **center of pressure** is the point at which the entire force could be concentrated to produce the same first moment with respect to the x-axis. Its y-coordinate, $\bar{y}$, is given by

$$M_x = \bar{y} \cdot F.$$

Find the y-coordinate of the center of pressure.

6. What did you learn as a result of doing this Exploration that you did not know before? (Over)

Exploration 78: Spindletop Oil Well Problem

Objective: Estimate the amount of oil that flowed from the Spindletop gusher in 1901.

On January 10, 1901 (the 10th day of the 20th century), as a culmination of work by Pattillo Higgins and Anthony Lucas, the first oil well gusher blew at a place called Spindletop close to Beaumont, Texas. The flow from the well was estimated initially to be 100,000 barrels per day, more than half the entire oil production of all other wells in the United States at that time. The well gushed for 9 days before it was finally controlled. In this Exploration you will make assumptions about the flow rate as a function of time, then use the assumptions to estimate the total amount of oil that flowed from the well in the 9-day period.

1. Assume that the flow rate decreased exponentially with time, dropping to half its original value at the end of the 9 days. Write an equation expressing barrels per day as a function of time.

2. By appropriate calculus, find the total number of barrels of oil that flowed from the well in the 9 days.

3. The oil that gushed from Spindletop in those 9 days was contained by building lakes using earthen dams. But most of it was lost. At today's oil prices, how much would that oil have been worth?

4. A barrel of oil is 42.5 gallons. A railroad tank car 60 feet long can hold 44,000 gallons of oil. How long a train of tank cars would it take to hold all the oil that flowed from Spindletop those 9 days?

5. For a fascinating account of the events preceding and following the Spindletop gusher, see *Pattillo Higgins and the Search for Texas Oil,* by Robert W. McDaniel with Henry C. Dethloff, Texas University Press, 1989.

6. What did you learn as a result of doing this Exploration that you did not know before?

Calculus Explorations
© 1998 Key Curriculum Press

Exploration 79: Two Geometric Series

Objective: Use geometric series as mathematical models of real-world phenomena that vary discretely rather than continuously.

Drug Dosage Problem: A patient takes a 30 mg (milligram) antibiotic capsule every hour. At the end of any one hour, the amount of antibiotic remaining in her body is only 90% of the amount at the beginning of that hour. Your objective is to predict the total amount in her body after many hours.

1. The first 30 mg dose is taken at time $t = 0$ hours. How much of this dose remains at the end of 1 hour? 2 hours? 3 hours? 4 hours?

2. Starting with the last dose, the amounts remaining from the first 5 doses ($t = 4$ hours) can be written

   $30, 30(0.9), 30(0.9)^2, 30(0.9)^3, 30(0.9)^4$.

   These numbers are part of a **geometric sequence**. The next term in the sequence is formed by multiplying the preceding term by 0.9, the **common ratio**. The total amount in the patient's body after these 5 doses is a **partial sum** of a **geometric series**,

   $30 + 30(0.9) + 30(0.9)^2 + 30(0.9)^3 + 30(0.9)^4$.

   How many milligrams of the antibiotic are in her body after these 5 doses?

3. The SUM and SEQUENCE commands on your grapher can be used to compute partial sums of any series. Write the appropriate commands to sum the formula $30(0.9)^n$ from $n = 0$ to $n = t$. Check your commands by showing that you get the same sum as in Problem 2 when $t = 4$ hours.

4. Find the 11th partial sum ($t = 10$) and the 21st partial sum ($t = 20$).

5. The partial sums for this series **converge** to a limit. What does this limit appear to be? What implications does the convergence of this series have for the patient in this problem?

Regular Savings Problem: Ernest Lee Dunn puts $800 into his IRA each year. The money earns 10% per year interest (APR), which means that at the end of any year the amount in the IRA is 1.1 times the amount at the beginning of that year.

6. The amount in the account at the beginning of year t is the sum of the worths of each deposit. Starting with the worth of the last deposit, the total is a partial sum of this geometric series:

   $800 + 800(1.1) + 800(1.1)^2 + \ldots + 800(1.1)^t$.

   Find the totals in the account when $t = 10, 20$, and 30 years.

7. Do the totals in Problem 6 seem to be **converging** to a finite limit or **diverging** toward infinity?

8. What did you learn as a result of doing this Exploration that you did not know before?

Exploration 80: A Power Series for a Familiar Function

Objective: Learn what a power series is, and how it can fit closely a particular function.

Let $P(x)$ be defined by

$$P(x) = 1 + x + \frac{1}{2!}x^2 + \frac{1}{3!}x^3 + \ldots \frac{1}{n!}x^n + \ldots$$

The letter P is used because the right side of the equation is a **power series.** It is also appropriate because the expression looks like a **polynomial,** except that it has an infinite number of terms. In this Exploration you will calculate and plot values of $P(x)$ and try to figure out which familiar function P represents.

1. Calculate $P(0.6)$ three times, using 3, 4, and 5 terms of the series ($n = 2, 3,$ and 4).

2. The values of $P(0.6)$ in Problem 1 are **partial sums** of the series. Use the SUM and SEQUENCE commands on your grapher to enter an equation into y_1 that will calculate $P(0.6)$ for $n = x$. Then make a table of values of $P(0.6)$ using $n = 5, 6, 7, 8, 9,$ and 10. What limit do the partial sums seem to be approaching as x increases?

3. Change the equation in y_1 so that it calculates the 11th partial sum ($n = 10$) of the series for $P(x)$. Then plot the graph using a friendly x-window of about $[-5, 5]$ and a y-window of $[-1, 10]$. Sketch the result here.

4. Find the 11th partial sum ($n = 10$) as in Problem 3 for $P(1)$. What familiar number does the answer resemble?

5. Make a conjecture about which one of the elementary functions on your grapher fits the series for $P(x)$ when x is not too far from zero. Give numerical or graphical evidence that your conjecture is correct.

6. Show that the 11th partial sum for this series is *not* close to the function you conjectured in Problem 5 at $x = 10$.

7. What did you learn as a result of doing this Exploration that you did not know before?

Calculus Explorations
© 1998 Key Curriculum Press

Exploration 81: Power Series for Other Familiar Functions

<u>Objective</u>: Given a power series in sigma notation, write out the first few terms, graph partial sums, and conjecture a target function.

1. The equation below specifies in **sigma notation** the power series for a familiar function.

$$f(x) = \sum_{n=0}^{\infty} (-1)^n \frac{1}{(2n+1)!} x^{2n+1}$$

Write out the first few terms of the power series. Then plot the fifth partial sums of the series ($n = 4$) in a friendly x-window of about $[-5, 5]$. Sketch the result here. Make a conjecture about which one of the elementary functions is the **target function** (i.e., fits well when x is close to zero).

2. For the series for $f(x)$ in Problem 1, find $f(0)$ and the first four derivatives, $f'(0), f''(0), f'''(0)$, and $f^{(4)}(0)$. Then find sin 0 and the first four derivatives of sin x evaluated at $x = 0$. What do you notice about these values? Give other evidence that sin x is the target function.

3. The equation below specifies in sigma notation the power series for a familiar function.

$$g(x) = \sum_{n=0}^{\infty} (-1)^n \frac{1}{(2n)!} x^{2n}$$

Write out the first few terms of the power series. Then plot the fifth partial sums of the series ($n = 4$) in a friendly x-window of about $[-5, 5]$. Sketch the result here. Make a conjecture about which one of the elementary functions is the target function.

4. The equation below specifies in sigma notation the power series for a familiar function.

$$h(x) = \sum_{n=0}^{\infty} \frac{1}{(2n)!} x^{2n}$$

Write out the first few terms of the power series. Then plot the fifth partial sums of the series ($n = 4$) in a friendly x-window of about $[-5, 5]$. Sketch the result here. Make a conjecture about which one of the elementary functions is the target function.

5. What did you learn as a result of doing this Exploration that you did not know before? (Over)

Exploration 82: A Power Series for a Definite Integral

<u>Objective</u>: Write a power series for the integrand in a definite integral, then do the integrating.

The figure shows the solid formed by rotating around the y-axis the region under the graph of $y = x \sin x$ from $x = 0$ to $x = \pi$. In this Exploration you will find the volume of the solid several ways.

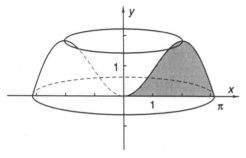

1. Using cylindrical shells, write an integral representing the volume of this solid.

2. The integrand in Problem 1 involves the expression $x^2 \sin x$. Write this expression as a Maclaurin series, taking advantage of the fact that you already know the series for $\sin x$.

3. Substitute the series in Problem 2 into the integral for the volume. Do the integrating (algebraically) and substitute the limits of integration. Use the sixth partial sum ($n = 5$) of the resulting series to find an estimate of the volume.

4. Evaluate the integral in Problem 1 directly, using the fundamental theorem. How close does the estimate in Problem 3 come to the actual volume?

5. What did you learn as a result of doing this Exploration that you did not know before?

Calculus Explorations
© 1998 Key Curriculum Press

Exploration 83: Introduction to the Ratio Technique

<u>Objective</u>: Find an interval of values of x for which the Taylor series for ln x converges.

1. Write the first few terms of the Taylor series for
   ln x expanded about $x = 1$.

2. Following the pattern in the series above, write the
   first nine terms of the series for ln 1.6.

n	t_n	
1		
2		
3		
4		
5		
6		
7		
8		
9		

3. Put another column in the table for Problem 2 that
   shows the absolute value of the ratio of successive
   terms, $|t_{n+1}/t_n|$.

4. Do the ratios of terms seem to be approaching a limit
   as n becomes large? If so, what does this limit seem
   to be? If not, tell why not.

5. If you do not simplify the values of t_n in Problem 2,
   a pattern shows up in the ratios that allows you to
   answer the question in Problem 4 very easily. Find
   this pattern.

6. Find the fourth partial sum of the series for ln 1.6.

7. Write the absolute values of the first five terms of the
   tail of the series after S_4.

8. Write the first five terms of the geometric series with
   first term $|t_5|$ and common ratio 0.7.

9. To what number does the geometric series in Problem
   8 converge?

10. Explain why the number in Problem 9 is an upper
    bound for the sum of the absolute values of the terms
    in the tail of the ln 1.6 series.

11. Explain why the technique of this exercise could
    not be used to find an upper bound for the tail of the
    ln series if 4 were substituted for x.

(Over)

Exploration 83, continued

12. Write the formula for t_n in the Taylor series for $\ln x$ expanded about $x = 1$. Then write a formula for the absolute value of the ratio of terms,

$$\left| \frac{t_{n+1}}{t_n} \right| .$$

15. The interval you found in Problem 14 is called the **interval of convergence.** The distance from the middle of this interval to one of its endpoints is called the **radius of convergence.** What is the radius of convergence of the Taylor series for $\ln x$ expanded about $x = 1$?

13. Find the limit of the fraction in Problem 12 as n approaches infinity. You should realize that x is a constant with respect to n and can be treated as such when you take the limit.

16. What did you learn as a result of doing this Exploration that you did not know before?

14. If the limit, L, in Problem 13 is less than 1, you can always find a convergent geometric series with common ratio between L and 1 that forms an upper bound for the tail of the ln series. By appropriate algebra, find an interval of x-values for which the ln series will converge if x is in this interval.

Name _____ Group _____ Date _____

Exploration 84: Improper Integrals to Test for Convergence

<u>Objective</u>: Prove that a *p*-series converges by comparing the tail of the series with a convergent improper integral.

The *p*-series $1 + \frac{1}{4} + \frac{1}{9} + \frac{1}{16} + \ldots$ converges. To demonstrate that this is true, you could draw a graph as shown in the figure below. The areas of the rectangles equal the terms of the tail after the fourth partial sum. The **remainder,** R_4, equals $t_5 + t_6 + t_7 + \ldots$.

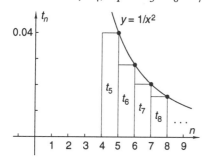

1. Explain why R_4 is a lower Riemann sum for the improper integral

$$\int_4^\infty 1/x^2 \, dx.$$

2. By evaluating the improper integral in Problem 1, find an upper bound for R_4.

3. Calculate the first, second, and third partial sums of the tail of the series starting at t_5.

4. Explain why the sequence of partial sums in Problem 3 is increasing, although the terms themselves are decreasing.

5. If a sequence is **increasing** and **bounded above,** then the sequence converges. How does this property allow you to conclude that the *series* for R_4 converges?

6. Based on the results of Problem 5, explain why the entire *p*-series converges.

7. The rectangles representing the terms of R_4 could have been drawn to form an *upper* sum.

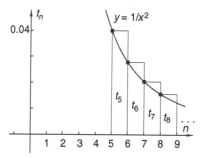

Explain why the improper integral from 5 to infinity would *not* be sufficient to prove that the tail of the series converges.

8. What did you learn as a result of doing this Exploration that you did not know before? (Over)

Calculus Explorations
© 1998 Key Curriculum Press

Exploration Masters / **89**

Exploration 85: Convergence of a Series of Constants

<u>Objective</u>: Determine whether or not various series of constants converge.

1. The **geometric series** $100 + 80 + 64 + \ldots$ can be shown to converge using the *definition* of convergence of a series. Demonstrate that you understand the definition by using it to prove that the series converges.

2. The **harmonic series** $\displaystyle\sum_{n=1}^{\infty} \frac{1}{n}$ diverges, even though the terms get smaller and approach 0 as a limit. Use an appropriate method to explain *why* the series diverges. A graph might help you do the explaining.

3. The Maclaurin series for cosh x is

$$\cosh x = 1 + \frac{1}{2!}x^2 + \frac{1}{4!}x^4 + \frac{1}{6!}x^6 + \ldots .$$

If you evaluate cosh 2 using this series you get

$$\cosh 2 = 1 + \frac{4}{2} + \frac{16}{24} + \frac{64}{720} + \ldots .$$

Write the fifth term of the series.

4. Explain why the partial sums of the series in Problem 3 are *increasing* although the terms of the series (after the second term) are *decreasing*.

5. The ratio t_3/t_2 equals $(64/720)/(16/24)$, which equals 4/30. Write the first few terms of a geometric series with first term 16/24 and common ratio 4/30. How do the terms of the geometric series compare with the terms of the cosh 2 series?

6. To what number does the geometric series in Problem 5 converge? How does the convergence of the geometric series imply convergence of the cosh 2 series?

7. What did you learn as a result of doing this Exploration that you did not know before? (Over)

Calculus Explorations
© 1998 Key Curriculum Press

Exploration 86: Introduction to Error Analysis for Series

Objective: Estimate the error made in using a partial sum of a series to approximate a function value.

1. Write the first three terms of the Maclaurin series for cos 0.6. Find the third partial sum of the series, S_2.

2. Calculate **error** in using S_2 to approximate cos 0.6 by subtracting cos $0.6 - S_2$, where the value of cos 0.6 is found on your calculator.

3. Show that the error you calculated in Problem 2 is less in absolute value than $|t_3|$, where t_3 is the first term of the tail of the series.

4. The error you calculated in Problem 2 is equal to another quantity you have computed for series. What is the name of this quantity?

5. Since the terms in the cos 0.6 series are strictly alternating, strictly decreasing in absolute value, and approach zero for a limit as n approaches infinity, the error in using S_n as an approximation for the limit of the series is no larger in absolute value than $|t_{n+1}|$. Use this information to find out how many terms of the series for cos 0.6 you would have to use in order for the partial sum to be correct to 20 decimal places (error $< 0.5 \times 10^{-20}$).

6. Write the first three terms of the Maclaurin series for $e^{0.6}$. Find S_2, the third partial sum of the series.

7. Calculate the error in using S_2 to estimate $e^{0.6}$.

8. The error in using S_2 to estimate $e^{0.6}$ is more than t_3, the first term of the tail, since all terms in the tail are positive. By what factor, F, is the error larger than t_3?

9. Show that the factor you calculated in Problem 8 is less than the maximum value of y''' on the interval $[0, 0.6]$, where $y = e^x$.

10. Look up the Lagrange form of the remainder of a Maclaurin series. How does the work in Problems 6 through 9 relate to what you find in the text?

11. What did you learn as a result of doing this Exploration that you did not know before?

Exploration 87: Error Analysis by Improper Integral

<u>Objective</u>: Use an improper integral to find upper and lower bounds for the remainder of a convergent p-series.

1. Find the fourth partial sum, S_4, of the p-series

$$\sum_{n=1}^{\infty} \frac{1}{n^2} .$$

2. You can tell that the series in Problem 1 converges because the exponent of n is greater than 1. To find an estimate of the **error** in using S_4 to estimate the number to which the series converges, you can find an upper bound using an improper integral. The figure below shows that

$$y = \frac{1}{(x-1)^2}$$

is an upper bound for the terms of the series.

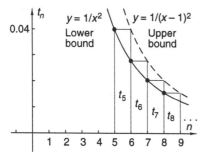

By evaluating an appropriate improper integral, find an upper bound for R_4, the remainder starting at term t_5.

3. The graph also shows that

$$y = \frac{1}{x^2}$$

is a *lower* bound for the terms of the series. By evaluating another improper integral, find a lower bound for R_4.

4. Based on your answer to Problem 3, is the fourth partial sum of this p-series a very good estimate of the limit to which the series converges? Explain.

5. Find S_{100} for this p-series. Find upper and lower bounds for the remainder. Would there be very much error in using S_{100} as an estimate for the limit to which this p-series converges? What is your best estimate of the limit?

6. What did you learn as a result of doing this Exploration that you did not know before? (Over)

Calculus Explorations
© 1998 Key Curriculum Press

Solutions for the Explorations

Limits, Derivatives, Integrals, and Integrals

Exploration 1: Instantaneous Rate of Change of a Function

1. Such a graph might look like this:

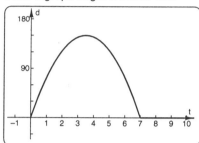

2. $d = 200t \cdot 2^{-t}$:

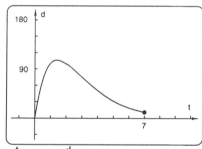

3.
t	d
0	0.0°
1	100.0°
2	100.0°
3	75.0°
4	50.0°
5	31.3°
6	18.8°
7	10.9°
8	6.3°
9	3.5°
10	2.0°

4. Door appears to be opening. The graph of d shows that d was less than 100° before t = 1 sec, and greater than 100° after t = 1 sec.
5. Average Rate = (change in value) / (Time)
    $= (200(1.1) \cdot 2^{-1.1} - 200(1) \cdot 2^{-1}) / (1.1 - 1)$
    $\approx (102.6° - 100°) / 0.1$ sec
    $= 26°$ / sec.
    This number is greater than zero, which seems to show that the door is still opening because d is increasing.
6. Average rate for time interval [1, 1.01]
    $\approx 30°$ / sec.
    Average rate for time interval [1, 1.001]
    $\approx 31°$ / sec.

Average rate for time interval [1, 1.0000001]
$\approx 31°$ / sec.
The average rate seems to be approaching
30.68° / sec ≈ 31° sec!
7. [Paragraph telling what you have learned.]

Exploration 2: Graphs of Functions

1. a. $f(x) = 3^{-x}$:

    b. Decreasing slowly.
2. a. $f(x) = \sin \frac{\pi}{2} x$:

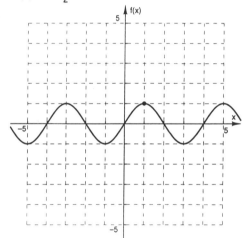

    b. Not changing.

3. a. $f(x) = x^2 + 2x - 2$:

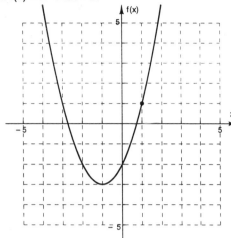

b. Increasing quickly.

4. a. $f(x) = \sec x$:

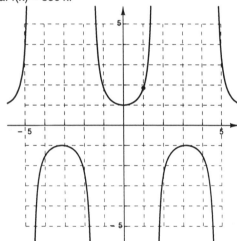

b. Increasing quickly.

5. a. $f(x) = \dfrac{1}{x}$:

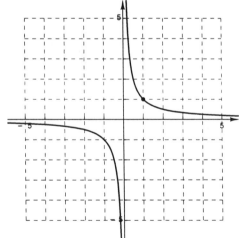

b. Decreasing slowly.

Exploration 3: Introduction to Definite Integrals

1. From $t = 30$ and $t = 50$ seconds, the velocity seems to be about 60 ft/sec.
   Distance = rate · time, so the distance traveled is about 60 ft/sec · (50 sec − 30 sec) = 1200 ft.

2. The rectangle on the graph with height = 60 and base from $t = 30$ to $t = 50$ has area = base · height = 1200.

3. The sample partial square has about 0.6 square spaces under the curve. All the partial squares under the graph from $t = 0$ to $t = 20$ have area about 28.6 square spaces.

4. Each small space has base representing 5 sec, and height representing 10 ft/sec. So the area of each small square = base · height = 5 sec · 10 ft/sec represents 50 ft. Therefore, the distance was about 28.6 · 50 = 1430 ft.
   (Eqn. is $v(t) = 60 + 40(0.92^{2x})$. Precise answer is 1431.3207. . . ft.)

5. The x-value is in seconds, and the y-value is in feet/second, so their product (i.e., the definite integral) is in seconds · feet/second = feet.

6. The squares and partial squares under the curve have about 45.2 square spaces of area. Each square space has base representing 1, and height representing 5. So one square space represents 5 units of definite integral, and the total definite integral is about 226 units.
   (Eqn. is $y = (\pi/4)[36 - (x - 6)]^2$. Precise answer is 226.1946. . . .)

7. The x-units are in inches, and the y-units are in square inches. So their product, the definite integral is in *cubic* inches. The definite integral seems to represent the *volume* of the football.

8. [Learn]

Exploration 4: Definite Integrals by Trapezoidal Rule

1. The graph shows time on the x-axis and velocity on the y-axis. A *definite integral* can be used to find time · velocity = distance.

2. There are about 60.8 square spaces under the graph. Each square represents about 25 miles, so the total distance is about 1520 miles.

3. $v(t) = t^3 - 21t^2 + 100t + 110$
   The graph confirms Ella's conclusion. Tracing to integer values of t shows the same v(t) values as on the graph.

4. Divide the graph into "trapezoids" of width = 2.

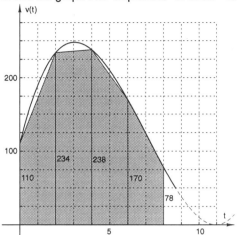

Areas of trapezoids are

$\frac{1}{2}(110 + 234) \cdot 2 = 344$

$\frac{1}{2}(234 + 238) \cdot 2 = 472$

$\frac{1}{2}(238 + 170) \cdot 2 = 408$

$\frac{1}{2}(170 + 78) \cdot 2 = 248$

Integral ≈ sum = 1472, which is reasonably close to the area found in Problem 2.

5. Sample program for TI-82 TRAPRULE

```
Prompt A          X+D→X
Prompt B          Y₁+S→S
Prompt N          Disp Y₁
(B-A)/N→D         End
0→S               B→X
A→X               Y₁/2+S→S
Y₁/2→S            Disp SD
For(K,1,N-1,1)
```

6. Using 20 trapezoids, the definite integral is about 1518.08 miles.
7. The approximate definite integral is:
     1519.52 miles for 40 trapezoids
     1519.9232 miles for 100 trapezoids
     1519.999232 miles for 1000 trapezoids
   The exact definite integral appears to be 1520 miles!
8. According to the graph, Ella's greatest velocity was about 248 miles per minute at about t = 3 minutes.

   (Actual maximum was $124 + \frac{94}{3} \cdot \sqrt{\frac{47}{3}}$ miles, at

   $t = 7 - \sqrt{\frac{47}{3}}$ minutes.)
9. v(4.9) = 213.439, and v(5.1) = 206.441.
   Velocity changed −6.998 ft/sec in 0.2 sec.
   Rate of change ≈ −6.998/0.02 = −34.99.
   So Ella was slowing down at about 35 miles/minute per minute.
10. Ella stops gradually at t = 11. The graph gently levels off to v(t) = 0.
11. [Learn]

Exploration 5: Introduction to Limits

1. $f(x) = \dfrac{x^3 - 7x^2 + 17x - 15}{x - 3}$
   Graph, showing a gap at x = 3, but that f(x) is close to 2 when x is close to, but not equal to, 3.

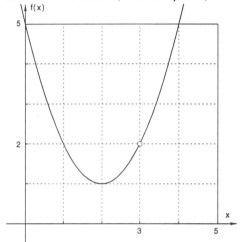

2. $f(3) = \dfrac{27 - 63 + 51 - 15}{3 - 3} = \dfrac{0}{0}$,
   an indeterminate form.
3. f(x) is very close to 2 when x is close to 3. The limit is 2.
4.

x	f(x)
2.5	1.25
2.6	1.36
2.7	1.49
2.8	1.64
2.9	1.81
3.0	?.?? (does not exist)
3.1	2.21
3.2	2.44
3.3	2.69
3.4	2.96
3.5	3.25

5. f(x) stays between 1.25 and 3.25 if x is between 2.5 and 3.5 (and x ≠ 3).
6. $f(x) = \dfrac{x^3 - 7x^2 + 17x - 15}{x - 3} = x^2 - 4x + 5, x \neq 3$.

   Set $x^2 - 4x + 5 = 1.99$ and solve algebraically or numerically. Repeat with $x^2 - 4x + 5 = 2.01$.
   If 2.99498743. . . < x < 3.00498756. . ., then f(x) will be between 1.99 and 2.01. (Also if x is between 0.9950. . . and 1.0050. . ., but this is not the interval of interest!)
7. On the left, keep x within 0.0050125. . . unit of 3.
   On the right, keep x within 0.0049875. . . unit of 3.
8. "If x is within 0.0049875... unit of 3 (but not equal to 3), then f(x) is within 0.01 unit of 2."
9. See the text definition.
10. L = 2, c = 3, ε = 0.01, δ = 0.0049. . . .
11. [Learn]

Exploration 6: The Definition of Limit

1. For any number $\varepsilon > 0$, there is a number $\delta > 0$ such that if x is within δ units of 5 (but $x \neq 5$), then f(x) is within ε units of 6.
2. f(x) is less than $6 + 1 = 7$ if x is greater than about 2.6, so x must be kept within about 2.4 units of 5 on the left.
3. f(x) is greater than $6 - 1 = 5$ if x is less than about 5.9, so x must be kept within about 0.9 units of 5 on the right.
4. If δ is about 0.9, then f(x) will be within 1 unit from 6 whenever x is within δ units of 5.
5. Set f(x) = 5.
$$4 - 2(x - 6)^{1/3} = 5$$
$$(x - 6)^{1/3} = -0.5$$
$$x - 6 = -0.125$$
$$|x - 5| = 0.875$$

   Set $\delta = 0.875$.
6. If f(x) = 6.01,
$$4 - 2(x - 6)^{1/3} = 6.01$$
$$(x - 6)^{1/3} = -1.005$$
$$x - 6 = -1.015075125$$
$$|x - 5| = 0.015075125$$
   If f(x) = 5.99,
$$4 - 2(x - 6)^{1/3} = 5.99$$
$$(x - 6)^{1/3} = -0.995$$
$$x - 6 = -0.985074875$$
$$|x - 5| = 0.014925125$$

   Set $\delta = 0.014925125$.
7. If f(x) = 6.0001,
$$4 - 2(x - 6)^{1/3} = 6.0001$$
$$(x - 6)^{1/3} = -1.00005$$
$$x - 6 = -1.000150007500125$$
$$|x - 5| = 0.000150007500125$$
   If f(x) = 5.9999,
$$4 - 2(x - 6)^{1/3} = 5.9999$$
$$(x - 6)^{1/3} = -0.99995$$
$$x - 6 = -0.999850007499875$$
$$|x - 5| = 0.000149992500125$$

   Set $\delta = 0.000149992500125$.
8. If f(x) = $6 + \varepsilon$,
$$4 - 2(x - 6)^{1/3} = 6 + \varepsilon$$
$$(x - 6)^{1/3} = -\left(1 + \tfrac{1}{2}\varepsilon\right)$$
$$x - 6 = -\left(1 + \tfrac{1}{2}\varepsilon\right)^3$$
$$|x - 5| = \left(1 + \tfrac{1}{2}\varepsilon\right)^3 - 1$$
$$= \tfrac{3}{2}\varepsilon + \tfrac{3}{4}\varepsilon^2 + \tfrac{1}{8}\varepsilon^3$$
   If f(x) = $6 - \varepsilon$,
$$4 - 2(x - 6)^{1/3} = 6 - \varepsilon$$
$$(x - 6)^{1/3} = -\left(1 - \tfrac{1}{2}\varepsilon\right)$$
$$x - 6 = -\left(1 - \tfrac{1}{2}\varepsilon\right)^3$$
$$|x - 5| = 1 - \left(1 - \tfrac{1}{2}\varepsilon\right)^3$$

$$= \tfrac{3}{2}\varepsilon - \tfrac{3}{4}\varepsilon^2 + \tfrac{1}{8}\varepsilon^3$$
   Set $\delta = 1 - \left(1 - \tfrac{1}{2}\varepsilon\right)^3$
9. [Learn]

Exploration 7: Extension of the Limit Theorems by Mathematical Induction

1. For n = 2, $f_2(x) = g_1(x) + g_2(x)$.
   By the limit of a sum (of two functions) property,
$$\lim_{x \to c} f_2(x) = \lim_{x \to c} g_1(x) + \lim_{x \to c} g_2(x) = L_1 + L_2.$$
2. Assume the property is false.
   Then there is a number n = j such that
$$\lim_{x \to c} f_j(x) \neq L_1 + \ldots + L_j.$$
3. 2 is in T and j is in F.

4. $f_3(x) = g_1(x) + g_2(x) + g_3(x)$
   $= g_1(x) + [g_2(x) + g_3(x)]$
$$\therefore \lim_{x \to c} f_3(x) = \lim_{x \to c} g_1(x) + \lim_{x \to c} [g_2(x) + g_3(x)]$$
   by the limit of a sum of two functions
$$= \lim_{x \to c} g_1(x) + \lim_{x \to c} g_2(x) + \lim_{x \to c} g_3(x)$$
   by the limit of a sum of two functions again
   $= L_1 + L_2 + L_3$.

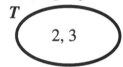 

5. Assume that $\lim_{x \to c} f_6(x) =$
   $f_6(x) = g_1(x) + g_2(x) + g_3(x) + g_4(x) + g_5(x) + g_6(x)$.
   If the property is true for n = 5, then
$$\lim_{x \to c} f_5(x) = L_1 + L_2 + L_3 + L_4 + L_5.$$
   Then $\lim_{x \to c} f_6(x)$
$$= \lim_{x \to c} (g_1(x) + g_2(x) + g_3(x) + g_4(x) + g_5(x) + g_6(x))$$
$$= \lim_{x \to c} [g_1(x) + g_2(x) + g_3(x) + g_4(x) + g_5(x)]$$
$$+ \lim_{x \to c} g_6(x)$$
   by the limit of a sum of two functions
$$= \lim_{x \to c} f_5(x) + \lim_{x \to c} g_6(x) \text{ by the definition of } f_5(x)$$
   $= L_1 + L_2 + L_3 + L_4 + L_5 + L_6$ by the assumption about $f_5(x)$ and the definition of L_6
   Thus the property is true for n = 6, Q.E.D.
6. The set is nonempty because $j \in \boldsymbol{F}$ (by assumption in Problems 2 and 3 above).
   The elements of F are all positive since each element is greater than 2.

7. If $I \in F$, then $I > 2$, so $I - 1 > 1$ is positive. $I - 1$ cannot be in F, because I is the smallest element of F, and $I - 1$ is smaller than I.

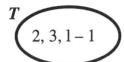 

T : $2, 3, I - 1$ F : I, j

8. $\lim\limits_{x \to c} f_{I-1}(x) = L_1 + \ldots + L_{I-1}$ because $I - 1 \in T$.

$\lim\limits_{x \to c} f_I(x) \neq L_1 + \ldots + L_{I-1} + L_I$ because $I \in F$.

9. $f_I(x) = g_1(x) + \ldots + g_{I-1}(x) + g_I(x)$.
$= f_{I-1}(x) + g_I(x)$

10. $\lim\limits_{x \to c} f_I(x) = \lim\limits_{x \to c} f_{I-1}(x) + \lim\limits_{x \to c} g_I(x)$
$= \lim\limits_{x \to c} f_{I-1}(x) + L_I$

11. $\lim\limits_{x \to c} f_I(x) = (L_1 + \ldots + L_{I-1}) + L_I$
$= L_1 + \ldots + L_{I-1} + L_I$

12. In Problem 8, $\lim\limits_{x \to c} f_I(x) \neq L_1 + \ldots + L_{I-1} + L_I$, but in Problem 11, $\lim\limits_{x \to c} f_I(x) = L_1 + \ldots + L_{I-1} + L_I$!

13. The assumption (property is false for some integers $n \geq 2$) must be false. Thus there are *no* positive integers, n, for which the property is false. Thus the property is true for all integers $n \geq 2$.

Proof:
Anchor:
The property is true for n = 2 (Problem 1, above)
Induction hypothesis:
Assume the property is true for n = k > 2,
so $\lim\limits_{x \to c} f_k(x) = L_1 + \ldots + L_k$.
Verification for n = k + 1:
$f_{k+1}(x) = g_1(x) + \ldots + g_k(x) + g_{k+1}(x)$
$= f_k(x) + g_{k+1}(x)$
$\lim\limits_{x \to c} f_{k+1}(x) = \lim\limits_{x \to c} f_k(x) + \lim\limits_{x \to c} g_{k+1}(x)$, because the property is true for n = 2.
$\lim\limits_{x \to c} f_{k+1}(x) = (L_1 + \ldots + L_k) + L_{k+1}$ using the induction hypothesis.
The property is true for n = k + 1.
Conclusion:
The property is true for *all* $n \geq 2$, Q.E.D.

14. [Learn]

Exploration 8: Continuous and Discontinuous Functions

1. $f(x) = \begin{cases} x + 1, & \text{if } x < 2 \\ (x - 5)^2, & \text{if } x \geq 2 \end{cases}$
Graph, showing a step discontinuity at x = 2.

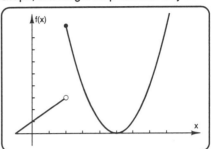

2. A function is discontinuous at a point if the function's value does not equal its limit, or if the limit does not exist, or if the function is undefined at that point.

3. $\lim\limits_{x \to 2^-} f(x) = \lim\limits_{x \to 2^-} (x + 1) = 3$
$\lim\limits_{x \to 2^+} f(x) = \lim\limits_{x \to 2^+} k(x - 5)^2 = 9k$
The two limits must be equal for f to be continuous.

4. $9k = 3 \Rightarrow k = \frac{1}{3}$
$f(x) = \begin{cases} x + 1, & \text{if } x < 2 \\ \frac{1}{3}(x - 5)^2, & \text{if } x \geq 2 \end{cases}$
Graph, showing continuity at x = 2.

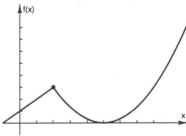

5. *Cusp* comes from the Latin *cuspis,* meaning a point or a pointed end. The word *bicuspid* for a tooth with two points comes from the same root. The word is appropriate for this graph since it comes to a point at x = 2.

6. At x = 2, f is neither increasing nor decreasing. The derivative is undefined since the backward difference quotients approach 1 as x approaches 2 from the left, and the forward difference quotients approach −2 as x approaches 2 from the right. There is no *one* number the derivative approaches.

7. [Learn]

Exploration 9: Limits Involving Infinity

1. Graph, $d(t) = 50 + 30(0.9)^t \cdot \cos \pi t$.

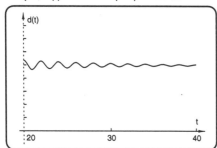

2. As t becomes very large, $d(t)$ stays close to 50.
3. Since $\cos \pi t$ varies between -1 and 1, $d(t)$ will be within $\varepsilon = 0.1$ of the limit whenever $30(0.9)^t$ is less than 0.1.

   Let $30(0.9)^t = 0.1$

   $0.9^t = 0.1/30$

   $t = \dfrac{\log(0.1/30)}{\log 0.9} = 54.1351\ldots$

   Making $t > 54.1351\ldots$ will guarantee that $d(t)$ is within 0.01 of 50.

4. After a long time, the pendulum moves very little and stays around 50 cm from the wall.
5. False. For example, as t gets larger from 0.5 to 1, $d(t)$ is getting farther away from the limit. However, it is true that as t gets larger, the maximum possible distance from $d(t)$ to the limit decreases, so that $d(t)$ remains within a closer distance to L.
6. See the text definition.
7. $f(x)$ may not actually "approach" L, as the example of this exploration shows. However, $f(x)$ "remains close" to L.
8. [Learn]

Exploration 10: Partial Rehearsal for Test on Limits

1. $\displaystyle\lim_{x \to 2} f(x) = \lim_{x \to 2} x^3 = 8$.

   Make $f(x)$ within 0.1 unit of 8.

   $7.9 < x^3 < 8.1$

   $\sqrt[3]{7.9} < x < \sqrt[3]{8.1}$

   $1.99163170\ldots < x < 2.00829885\ldots$

   $-0.008368\ldots < x - 2 < 0.008298\ldots$

   Keep x within $\delta = 0.008298\ldots$ unit of 2.

   Make $f(x)$ within ε unit of 8.

   $\sqrt[3]{8-\varepsilon} < x < \sqrt[3]{8+\varepsilon}$

   Keep x within $\delta = \sqrt[3]{8+\varepsilon} - 2$.

   δ is positive for any $\varepsilon > 0$ since $\sqrt[3]{8+\varepsilon} > 2$.

   Since there is a positive number δ for any positive value of ε such that keeping x within δ unit of 2 keeps $f(x)$ within ε unit of 8, 8 really is the limit of $f(x)$ as x approaches 2.

2. $g(0.5) = \sin \dfrac{\pi}{2} = 1$.

   Graph, $g(x) = \sin \pi x$.

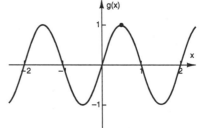

Make $g(x)$ within 0.01 of 1.

$0.99 < \sin \pi x < 1.01$

Since $\sin \pi x$ is bounded above by 1, only the left inequality is meaningful.

$\dfrac{1}{\pi} \sin^{-1} 0.99 < x < \dfrac{1}{\pi}(\pi - \sin^{-1} 0.99)$

$0.4549\ldots < x < 0.5450\ldots$

Keep x within $\delta = 0.0450\ldots$ unit of 0.5.

Make $g(x)$ within ε unit of 1.

Keep x within $\delta = 0.5 - \dfrac{1}{\pi} \sin^{-1}(1 - \varepsilon)$ of 0.5.

δ is positive when ε is small and positive.

3. $h(x) = 2^{x-3} + 5$

   $h(3) = 2^0 + 5 = 6$

   Make $h(x)$ within ε unit of 6.

   $6 - \varepsilon < 2^{x-3} + 5 < 6 + \varepsilon$

   $1 - \varepsilon < 2^{x-3} < 1 + \varepsilon$

   $\log_2(1 - \varepsilon) < x - 3 < \log_2(1 + \varepsilon)$

   Since $|\log_2(1 - \varepsilon)| > |\log_2(1 + \varepsilon)|$,

   make $\delta = \log_2(1 + \varepsilon)$.

4. $r(x) = \dfrac{\sin \pi x - 0.5}{x - 1/6}$

   $r(1/6)$ takes on the *indeterminate form* 0/0. Graph. Limit appears to be just below 3. (Exact value: $\pi \cos(\pi/6) = 2.7206\ldots$)

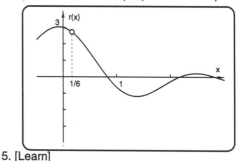

5. [Learn]

Exploration 11: One-Problem Summary of Calculus So Far

1. There are about 9 square spaces under the curve from $x = 1$ to $x = 7$. Each square space represents 10 units of area, so the definite integral is about 90.

2. Graph, showing tangent line at x = 1 with slope that appears to be about 1.5. Since each vertical space is 10 units and each horizontal space is 1 unit, the derivative is about 15.

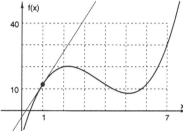

3. The graph is correct.
4. $f'(1) \approx \dfrac{f(1.01) - f(0.99)}{0.02} = 15.0001$
   This is very close to the estimate in Problem 2.
5. Using the trapezoidal rule with n = 6, the definite integral is about 91, which is reasonably close to the guess in Problem 1.
6. Using the trapezoidal rule with:
   n = 10—integral ≈ 90.36
   n = 100—integral ≈ 90.0036
   n = 1000—integral ≈ 90.000036
   The exact value appears to be 90.
7. In Problem 6, the exact value was found by taking the number of increments, n, to be very large. So a definition of the definite integral might be the *limit* of the trapezoidal rule value as n approaches infinity.
8. f(1) = 12, so
   $$m(x) = \frac{x^3 - 11x^2 + 34x - 12 - 12}{x - 1}$$
   $$= \frac{x^3 - 11x^2 + 34x - 24}{x - 1}$$

Graph:

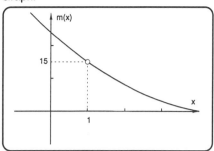

9. $m(x) = \dfrac{x^3 - 11x^2 + 34x - 24}{x - 1}$
   $$= \frac{(x - 1)(x^2 - 10x + 24)}{x - 1}$$
   $$= x^2 - 10x + 24 \quad (x \neq 1)$$
   $\displaystyle\lim_{x \to 1} m(x) = 15.$

10. $\displaystyle\lim_{x \to 1} m(x)$
    $= \displaystyle\lim_{x \to 1} \dfrac{f(x) - f(1)}{x - 1}$ Def.
    $= \displaystyle\lim_{x \to 1} \dfrac{x^3 - 11x^2 + 34x - 24}{x - 1}$ algebra
    $= \displaystyle\lim_{x \to 1} \dfrac{(x - 1)(x^2 - 10x + 24)}{x - 1}$ algebra
    $= \displaystyle\lim_{x \to 1} (x^2 - 10x + 24)$ since x ≠ 1 by def.
    of a limit
    $= \displaystyle\lim_{x \to 1} (x - 6)(x - 4)$ algebra
    $= \displaystyle\lim_{x \to 1} (x - 6) \lim_{x \to 1} (x - 4)$ Lim of a prod
    $= (^-5)(^-3)$ Lim of a linear fn 2x
    $= 15$ Q.E.D.

11. [Learn]

Derivatives, Antiderivatives, and Indefinite Integrals

Exploration 12: Exact Value of a Derivative

1. f(4) = 10, as shown on the graph.
2. $f'(4) = \displaystyle\lim_{x \to 4} \dfrac{f(x) - f(4)}{x - 4}$
3. $f'(4) = \displaystyle\lim_{x \to 4} \dfrac{x^3 - 4x^2 - 9x + 46 - 10}{x - 4}$
   $= \displaystyle\lim_{x \to 4} \dfrac{x^3 - 4x^2 - 9x + 36}{x - 4}$
   $= \displaystyle\lim_{x \to 4} \dfrac{(x - 4)(x^2 - 9)}{x - 4}$
   $= \displaystyle\lim_{x \to 4} (x^2 - 9)$
   $= 7$

4. Graph, showing that the line with slope 7 going through the point (4, 10) is tangent to the graph. (Equation of line is y = 7x − 18.)

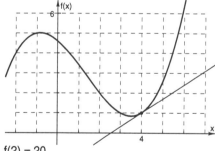

5. f(2) = 20
   $f'(2) = \displaystyle\lim_{x \to 2} \dfrac{f(x) - f(2)}{x - 2}$
   $= \displaystyle\lim_{x \to 2} \dfrac{x^3 - 4x^2 - 9x + 46 - 20}{x - 2}$
   $= \displaystyle\lim_{x \to 2} \dfrac{x^3 - 4x^2 - 9x + 26}{x - 2}$
   $= \displaystyle\lim_{x \to 2} \dfrac{(x - 2)(x^2 - 2x - 13)}{x - 2}$

$$= \lim_{x \to 2} x^2 - 2x - 13$$
$$= -13.$$

This appears reasonable because the function is decreasing at x = 2 (about twice as fast as it increases at x = 4), and the calculated value of the slope is negative (and in absolute value about twice the value of f'(4)).

6. [Learn]

Exploration 13: Numerical Derivative by Grapher

1. Graph, $d(t) = 200t \cdot 2^{-t}$, is correct.

2. $d'(t) = \dfrac{d(1.001) - d(0.999)}{0.002} = 30.6853\ldots$

3. Numerical derivative = 30.6853...
   Answer agrees with Problem 2.

4. $d'(2) \approx -19.3147\ldots$
   d'(1) is positive, which corresponds to the graph of d increasing at t = 1. d'(2) is negative, which corresponds to the graph of d decreasing at t = 2.

5. Graph, showing the numerical derivative. (Instructor check.)

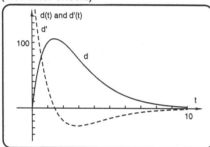

6. At the point where d'(t) = 0, the graph of d is horizontal. The door has just finished opening and is about to begin closing at this time.

7. d'(t) = 0 at t ≈ 1.4426...

8. Minimum is at x ≈ 2.8853...
   f(2.8853...) = 78.0991...

9. At the point in Problem 8, the graph changes from bending downward to bending upward.

10. [Learn]

Exploration 14: Algebraic Derivative of a Power Function

1. $f'(c) = \lim_{x \to c} \dfrac{f(x) - f(c)}{x - c}$
   $= \lim_{x \to c} \dfrac{x^5 - c^5}{x - c}$

2. $f'(c) = \lim_{x \to c} \dfrac{x^5 - c^5}{x - c}$
   $= \lim_{x \to c} \dfrac{(x - c)(x^4 + cx^3 + c^2x^2 + c^3x + c^4)}{x - c}$
   $= \lim_{x \to c} (x^4 + cx^3 + c^2x^2 + c^3x + c^4)$
   $= 5c^4$

3. $f'(3) = 5 \cdot 3^4 = 405$
   Numerical derivative ≈ 405.0001
   Formula seems to work.

4. Conjecture: For $f(x) = x^{10}$, $f'(x) = 10x^9$
   Test at c = 2:
   $10 \cdot 2^9 = 5120$
   Numerical derivative ≈ 5120.0154
   Formula seems to work.

5. The formula for the derivative of x^n seems to be $f'(x) = nx^{n-1}$, for any positive integer n!

6. [Learn]

Exploration 15: Deriving Velocity and Acceleration from Displacement Data

1. $x(t) = at^b$
   a = 2820.5148...
   b = 0.8222...
   The correlation coefficient is r = 0.9974....

2. Graph. Equation fits data reasonably well.

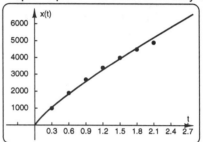

3. Using the regression equation for the displacement, $v(t) = x'(t) = b \cdot at^{b-1}$.

4. $v(0.9) \approx 0.8222\ldots \cdot 2820.5128\ldots (0.9)^{0.8222\ldots -1}$.
   = 2363.1354... ft/sec.

5. Velocity $\approx \dfrac{3400 - 1900}{1.2 - 0.6} = 2500$ ft/sec, which is reasonably close to the answer in problem 4.

6. Using the regression equation for the displacement: $a(t) = v'(t) = (b - 1) \cdot b \cdot at^{b-2}$.

7. x(1) = 2820.5148... ft
   v(1) = 2319.3025... ft/sec
   a(1) = -412.1456... (ft/sec)/sec

8. At 1 minute, x(60) ≈ 81752 ft (≈ 15.5 miles!). This does not seem reasonable; the data show the bullet to be slowing down more than the regression equation suggests.

9. [Learn]

Exploration 16: Derivative of the Sine of a Function

1. Graph, y = sin x.

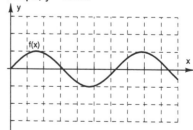

Calculus Explorations
© 1998 Key Curriculum Press

2. Graph, $y_2 = \cos x$, $y_3 =$ numerical derivative. The graphs are the same! The derivative really does seem to be cos x.

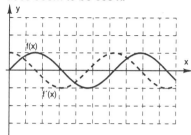

3. Graph, $g(x) = \sin 3x$

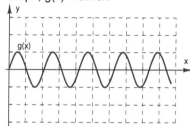

4. The graph is compressed horizontally. g(x) has 3 cycles for every 1 cycle of f(x).
5. The equation for g'(x) should be similar to the equation for f'(x) = cos x because g(x) is very similar to f(x).
   The graph of g'(x) should be "compressed" as g(x) is, so the equation should be like cos 3x. But because the graph of g(x) is compressed horizontally, all the slopes are increased. In other words, the slope is like Δy/Δx, so if Δx is compressed to a third, the slope is expanded triply.
   Conjecture: g'(x) = 3 cos 3x
   Graph, $y_2 = 3 \cos 3x$, $y_3 =$ numerical derivative.

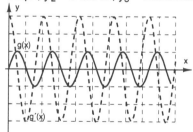

6. Graph, $h(x) = \sin x^2$ (solid).

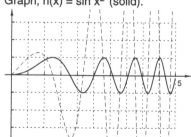

7. Conjecture: $h'(x) = 2x \cos x^2$
   Graph, at Problem 6 (dashed).

8. $t(x) = \sin x^{0.7}$
   Conjecture: $t'(x) = 0.7x^{-0.3} \cos x^{0.7}$
   Graph, verifying the conjecture.

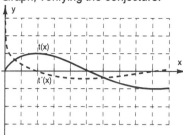

9. $f'(x) = g'(x) \cdot \cos [g(x)]$
10. [Learn]

Exploration 17: Rubber-Band Chain Rule Problem

1. Graph, F(x) versus x(t).

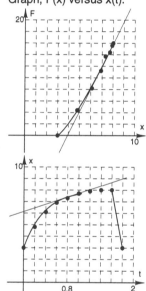

2. $\dfrac{dx}{dt}(0.8) \approx \dfrac{7.7 - 6.9 \text{ in}}{1.0 - 0.6 \text{ sec}} = 2$ in/sec
3. $\dfrac{dF}{dx}(7.3) \approx \dfrac{14.4 - 11.2 \text{ oz}}{7.7 - 6.9 \text{ in}} = 4$ oz/in
4. Graph, at Problem 1, showing that lines through the respective points with the slopes as found in Problems 2 and 3 are tangent to the graphs.
5. Graph, F(t).

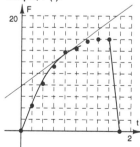

6. $\dfrac{dF}{dt} = \dfrac{dF}{dx} \cdot \dfrac{dx}{dt} = 4$ oz/in $\cdot$ 2 in/sec = 8 oz/sec

7. $\frac{dF}{dt}(0.8) = \frac{14.4 - 11.2 \text{ oz}}{1.0 - 0.6 \text{ sec}} = 8$ oz/sec — same answer as in Problem 6!

8. See graph at Problem 5. The line with slope 8 is tangent to the graph. (Observe the different scales for the two axes.)

9. [Learn]

====================

Exploration 18: Algebraic Derivative of Sine Problem

1. $y(-0.1) = 0.9983\dots$
   $y(0.1) = 0.9983\dots$
   $y(-0.01) = 0.999983\dots$
   $y(0.01) = 0.999983\dots$
   $y(-0.001) = 0.99999983\dots$
   $y(0.001) = 0.99999983\dots$
   y appears to approach 1 near x = 0.

2. $\sin 0.1 = 0.0998\dots < 0.1 < 0.1003\dots = \tan 0.1$
   $\sin 0.01 = 0.0099\dots < 0.01 < 0.0100\dots = \tan 0.01$
   $\sin 0.001 = 0.0009\dots < 0.001$
   $\quad\quad < 0.0010\dots = \tan 0.001$
   Statement appears to be true.

3. $\sin x < x < \tan x = \dfrac{\sin x}{\cos x}$

   Divide by sin x.

   $1 < \dfrac{x}{\sin x} < \dfrac{1}{\cos x} = \sec x$

4. Problem 3 says $1 < \dfrac{x}{\sin x} < \dfrac{1}{\cos x}$ as x approaches 0 from the right; therefore

   $1 > \dfrac{\sin x}{x} > \cos x$ as x approaches 0 from the right.

   Let $f(x) = \dfrac{\sin x}{x}$, g(x) = 1, and h(x) = cos x.

   Then f(x) is between g(x) and h(x) as x approaches 0 from the right;
   $\displaystyle\lim_{x\to 0^+} g(x) = \lim_{x\to 0^+} 1 = 1$; and
   $\displaystyle\lim_{x\to 0^+} h(x) = \lim_{x\to 0^+} \cos x = 1.$

   So by the squeeze theorem, $\displaystyle\lim_{x\to 0^+} f_2(x) = 1$ also.

5. Use h for Δx.

   Recall: $\sin A - \sin B = 2 \cos\frac{1}{2}(A + B) \sin\frac{1}{2}(A - B)$

   $\dfrac{\sin (x+h) - \sin x}{h}$

   $= \dfrac{2 \cos\frac{1}{2}(2x+h) \sin\frac{1}{2}h}{h}$

   $= \cos (x + \frac{1}{2}h) \cdot \dfrac{\sin\frac{1}{2}h}{\frac{1}{2}h}$

   As h approaches zero, the first factor approaches cos x and the second factor approaches 1.

   Thus $\dfrac{d}{dx}(\sin x) = \cos x$, Q.E.D.

6. A *lemma* is something you prove as an important step in proving something else. So if you want to prove that all (live) cats are warm-blooded, you might first prove the lemma that all cats are mammals, and then use the well-known properties of mammals to reach your conclusion.

7. [Learn]

====================

Exploration 19: Displacement and Acceleration from Velocity

1. If $d(t) = 50t + \dfrac{6}{1.6}t^{1.6} = 50t + 3.75t^{1.6}$,
   then $d'(t) = 50 + 6t^{0.6}$.

2. The derivative of a constant is zero, so
   $\dfrac{d}{dt}(d(t) + C) = d'(t) + \dfrac{d}{dt}C = d'(t)$

3. $50t + \dfrac{6}{1.6}t^{1.6} + C$ must equal 100 when t = 0.
   $0 + 0 + C = 100$, so C = 100.

4. The condition gives Ray's initial location.

5. Using $d(t) = 50t + \dfrac{6}{1.6}t^{1.6} + 100$,
   $d(10) = 749.2901\dots$ ft
   $d(20) = 1552.5632\dots$ ft

6. $a(t) = \dfrac{d}{dt}(50 + 6t^{0.6}) = 3.6t^{-0.4}$

7. $a(10) = 1.4331\dots \approx 1.43$ (ft/sec)/sec
   $a(20) = 1.0861\dots \approx 1.09$ (ft/sec)/sec
   At t = 10 the acceleration is 0.347..., or about 0.35 (ft/sec)/sec higher.

8. The truck's velocity is $v_T(t) = 50$ so its position is given by $d_T(t) = 50t + C$. The initial condition is $d_T(0) = 160$, so $d_T(t) = 50t + 160$.
   Solve $d(t) = d_T(t) + 70$ for t:
   $50t + \dfrac{6}{1.6}t^{1.6} + 100 = 50t + 160 + 70$
   $\dfrac{6}{1.6}t^{1.6} = 130$

   $t = \sqrt[1.6]{34.6666\dots} = 9.1715\dots \approx 9.2$ sec.

9. [Learn]

Calculus Explorations
© 1998 Key Curriculum Press

Products, Quotients, and Parametric Functions

Exploration 20: Derivative of a Product

1. $g(x) = x^7 \Rightarrow g'(x) = 7x^6$
   $h(x) = x^{11} \Rightarrow h'(x) = 11x^{10}$
2. $f(x) = g(x) \cdot h(x) = x^7 \cdot x^{11} = x^{18}$
   $f'(x) = 18x^{17}$
3. $g'(x) \cdot h'(x) = (7x^6)(11x^{10}) = 77x^{16} \neq 18x^{17} = f'(x)$
4. $18x^{17} = 7x^6 \cdot x^{11} + x^7 \cdot 11x^{10}$
5. $f'(x) = g'(x) \cdot h(x) + g(x) \cdot h'(x)$
6. Use $g(x) = x^2$, $h(x) = \sin x$.
   Then $g'(x) = 2x$, $h'(x) = \cos x$,
   $f'(x) = g'(x) \cdot h(x) + g(x) \cdot h'(x)$
   $\qquad = 2x \sin x + x^2 \cos x$
7. Graph, $f(x)$, its numerical derivative, and $f'(x)$ from Problem 6, showing that the numerical and algebraic derivatives are equivalent.

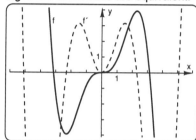

8. [Learn]

Exploration 21: Derivative of a Quotient— Do-It-Yourself!

1. $f(x) = \dfrac{x^3}{\sin x}$
   $f'(1) = 2.8021\ldots$ (numerically)
2. $\dfrac{d}{dx}(x^3) = 3x^2$, $\dfrac{d}{dx}(\sin x) = \cos x$
   $\dfrac{3(1)^2}{\cos 1} = 5.5524 \neq f'(1)$
3. $y + \Delta y = \dfrac{\Delta u + u}{\Delta v + v}$
4. $\dfrac{dy}{dx} = \lim\limits_{\Delta x \to 0} \dfrac{\Delta y}{\Delta x} = \lim\limits_{\Delta x \to 0} \dfrac{\dfrac{\Delta u + u}{\Delta v + v} - \dfrac{u}{v}}{\Delta x}$
5. $\dfrac{\Delta y}{\Delta x} = \dfrac{1}{\Delta x}\left(\dfrac{\Delta u + u}{\Delta v + v} - \dfrac{u}{v} \right)$
   $= \dfrac{1}{\Delta x}\left(\dfrac{(\Delta u + u)v}{(\Delta v + v)v} - \dfrac{u(\Delta v + v)}{v(\Delta v + v)} \right)$
   $= \dfrac{1}{\Delta x}\left(\dfrac{\Delta u \cdot v - u\Delta v}{v(\Delta v + v)} \right)$
6. $\dfrac{\Delta y}{\Delta x} = \dfrac{\dfrac{\Delta u}{\Delta x} \cdot v - u\dfrac{\Delta v}{\Delta x}}{v(\Delta v + v)}$
7. $\lim\limits_{\Delta x \to 0} \dfrac{\Delta y}{\Delta x} = \lim\limits_{\Delta x \to 0} \dfrac{\dfrac{\Delta u}{\Delta x} \cdot v - u\dfrac{\Delta v}{\Delta x}}{v(\Delta v + v)}$
   $= \dfrac{\lim\limits_{\Delta x \to 0} \dfrac{\Delta u}{\Delta x} \cdot v - \lim\limits_{\Delta x \to 0} u\dfrac{\Delta v}{\Delta x}}{\lim\limits_{\Delta x \to 0} v(\Delta v + v)}$

$= \dfrac{\dfrac{du}{dx} \cdot v - u\dfrac{dv}{dx}}{v^2} = \dfrac{u'v - uv'}{v^2}$

8. $y = x^3 \sin x$
   Use $u(x) = x^3$, $v(x) = \sin x$
   $\dfrac{dy}{dx} = \dfrac{u'v - uv'}{v^2}$
   $\qquad = \dfrac{3x^2 \sin x - x^3 \cos x}{(\sin x)^2}$
   $\dfrac{dy}{dx}(1) = \dfrac{3 \sin 1 - \cos 1}{(\sin 1)^2} = 2.8021\ldots$
   The numerical derivative is $2.8021\ldots$, which agrees with the algebraic derivative.
9. Take the derivative of the top function times the bottom function, minus the top function times the derivative of the bottom function, all divided by the bottom function squared.
10. [Learn]

Exploration 22: Derivatives of Inverse Trigonometric Functions

1. Graph, $y = \sin^{-1} x$.

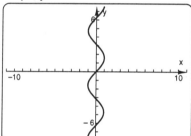

2. Graphs, $y = \sin x$ and $y = \sin^{-1} x$. The graphs are reflections of each other across the line $x = y$.

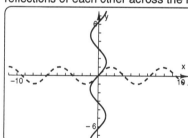

3. $x = \sin y$
   $1 = \cos y \cdot y'$, using the chain rule.
4. $y' = \dfrac{1}{\cos y} = \dfrac{1}{\cos (\sin^{-1} x)}$

5. Sketch:

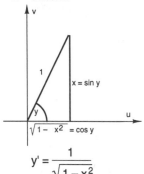

$$y' = \frac{1}{\sqrt{1-x^2}}$$

6. At $x = 0.5$, $y' = \dfrac{1}{\sqrt{1-0.25}} = 1.1547\ldots$

Graph. The line through $(0.5, \sin^{-1} 0.5)$ with slope $1.15\ldots$ is tangent to the graph.

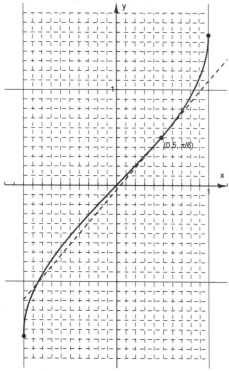

7. $y = \tan^{-1} x \Rightarrow x = \tan y$

$\Rightarrow 1 = (\sec y)^2 \cdot y'$

$\Rightarrow y' = \dfrac{1}{(\sec y)^2} = \dfrac{1}{1 + (\tan y)^2} = \dfrac{1}{1 + x^2}$

$y = \sec^{-1} x \Rightarrow x = \sec y$

$\Rightarrow 1 = \tan y \sec y \cdot y'$

$\Rightarrow y' = \dfrac{1}{\tan y \sec y} = \dfrac{1}{\sqrt{(\sec y)^2 - 1}\ \sec y}$

$\quad\ = \dfrac{1}{|x|\sqrt{x^2 - 1}}$

(See text for justification of $|x|$.)

8. [Learn]

Exploration 23: Differentiability Implies Continuity

1. $f(x)$ is continuous at $x = c$ if:
   $f(c)$ exists, $\displaystyle\lim_{x\to c} f(x)$ exists, and $\displaystyle\lim_{x\to c} f(x) = f(c)$.

2. $\displaystyle\lim_{x\to c} [f(x) - f(c)] = 0$ Given (note $f(c)$ exists)

   $\displaystyle\lim_{x\to c} f(c) = f(c)$ limit of a constant

   $\displaystyle\lim_{x\to c} [f(x) - f(c)] + \lim_{x\to c} f(c) = f(c)$ add equations

   $\displaystyle\lim_{x\to c} [f(x) - f(c) + f(c)] = f(c)$ sum of limits

   $\displaystyle\lim_{x\to c} f(x) = f(c)$

   Therefore, $f(x)$ is continuous at $x = c$.
   (See text for a slightly different set of algebraic steps.)

3. $f'(c) = \displaystyle\lim_{x\to c} \dfrac{f(x) - f(c)}{x - c}$

4. The limit of the denominator is zero.

5. $\displaystyle\lim_{x\to c} [f(x) - f(c)] = \lim_{x\to c} \left([f(x) - f(c)] \cdot \dfrac{x-c}{x-c} \right)$

   $\quad = \displaystyle\lim_{x\to c} \dfrac{f(x) - f(c)}{x - c} \cdot \lim_{x\to c} (x - c)$

   $\quad = f'(c) \cdot 0$

6. If f is differentiable at c, then $f'(c)$ is a real number
   $\Rightarrow f'(c) \cdot 0 = 0$ $f'(c)$ is a real number
   $\Rightarrow \displaystyle\lim_{x\to c} [f(x) - f(c)] = 0$ Problem 5
   $\Rightarrow f$ is continuous at $x = c$ Problem 2

7. The converse is: "If f is continuous at $x = c$, then f is differentiable at $x = c$."
   Counterexample: $f(x) = |x|$ is continuous at $x = 0$, but not differentiable at $x = 0$.

8. A *lemma* for a theorem is something you prove as an important step in proving the main theorem. An easy consequence of a previously proved result is called a *corollary*.

9. [Learn.]

Exploration 24: Parametric Function Graphs

1. $x = 0.4t \cos t$, $y = 0.3t + 2 \sin 2t$:

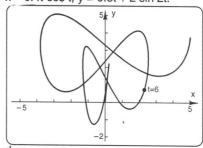

2. $\dfrac{dx}{dt} = 0.4 \cos t - 0.4t \sin t$

   $\dfrac{dy}{dt} = 0.3 + 4 \cos 2t$

   $x'(6) = 0.4 \cos 6 - 2.4 \sin 6 = 1.0546\ldots$
   $y'(6) = 0.3 + 4 \cos 12 = 3.6754\ldots$

Calculus Explorations
© 1998 Key Curriculum Press

3. $\dfrac{dy}{dx} = \dfrac{dy/dt}{dx/dt} = \dfrac{0.3 + 4\cos 2t}{0.4\cos t - 0.4t\sin t}$

$\dfrac{dy}{dx}(6) = \dfrac{3.6754\ldots}{1.0546\ldots} = 3.4849\ldots$ At $t = 6$, Adam is at
the point $(2.3044\ldots, 0.7268\ldots)$ on the graph. At
this point the graph is sloping upward with slope
considerably greater than 1. (Alternately: A line
through this point with slope $3.4849\ldots$ appears
tangent to the graph.)

4. dy/dx (numerically) equals $3.4849\ldots$, which agrees
   with the algebraic answer in Problem 3.

5. Adam turns around between about $t = 0.82$ and
   $t = 0.86$.

6. $y = 0$ at about 1.69:

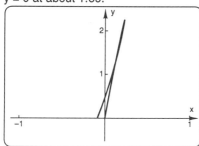

7. $y' = 0.3 + 4\cos 2t = 0$
   $\cos 2t = -0.075$
   $t = 0.5\cos^{-1} -0.075 = 0.8229\ldots$

8. At this time, Adam is not moving in the y-direction,
   but is still moving rightwards in the x-direction since
   $x'(0.8229\ldots) = 0.0306\ldots$

9. Graph, x-dilation of 50, y-dilation of 10.
   At $t = 0.8229\ldots$, the graph has a horizontal tangent
   so there is no cusp.

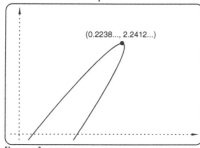

(0.2238..., 2.2412...)

10. [Learn]

Exploration 25: Implicit Relation Derivatives

1. Many values of x have more than one
   corresponding value of y.

2. $x^2 - 4xy + 4y^2 + x - 12y - 10 = 0$

   At $x = 6$, $36 - 24y + 4y^2 + 6 - 12y - 10 = 0$

   $4y^2 - 36y + 32 = 4(y - 1)(y - 8) = 0 \Rightarrow y = 1$ or 8.
   Graph, showing these points.

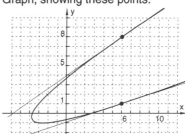

3. $2x - 4y - 4xy' + 8yy' + 1 - 12y' = 0$
   $2x - 4y + 1 = 4xy' - 8yy' + 12y'$

   $y' = \dfrac{2x - 4y + 1}{4x - 8y + 12} \left(= \dfrac{1}{2} - \dfrac{5}{4x - 8y + 12} \right)$

4. $y'(6,1) = \dfrac{12 - 4 + 1}{24 - 8 + 12} = \dfrac{9}{28}$

   $y'(6,8) = \dfrac{12 - 32 + 1}{24 - 64 + 12} = \dfrac{19}{28}$

   Graph, at Problem 2, showing that lines through
   (6, 1) and (6, 8) with slopes 9/28 and 19/28,
   respectively, are tangent to the graph.

5. The formula gives $y'(3, 1) = \dfrac{6 - 4 + 1}{12 - 8 + 12} = \dfrac{3}{16}$, but
   this has no meaning for the problem because (3, 1)
   does not even lie on the graph!

6. Rewrite the equation:
   $4y^2 - (4x + 12)y + (x^2 + x - 10) = 0$

   $y = \dfrac{4x + 12 \pm \sqrt{(4x + 12)^2 - 16(x^2 + x - 10)}}{8}$

   $= \dfrac{1}{2}\left(x + 3 \pm \sqrt{5x - 19}\right)$

   $y' = \dfrac{1}{2} \pm \dfrac{5}{4}(5x - 19)^{-1/2}$

7. For $Ax^2 + Bxy + Cy^2 + Dx + Ey + F = 0$, the
   discriminant is $B^2 - 4AC$. If the discriminant is
   positive, the graph is a hyperbola; if negative, an
   ellipse; and if zero, a parabola. For the given figure,
   $B^2 - 4AC = (-4)^2 - 4(1)(4) = 0$.
   Thus, the graph is a parabola, Q.E.D.

8. [Learn]

Definite and Indefinite Integrals

Exploration 26: A Motion Antiderivative Problem

1. $v(t) = 3t^{0.6}$

t	velocity	v(t)
0	0	0.0
2	4.5	4.5471...
4	6.9	6.8921...
6	8.8	8.7904...
8	10.4	10.4466...
10	11.9	11.9432...

   Thus the values of v(t) fit the data closely.

2. $a(t) = v'(t) = 1.8t^{-0.4}$
   a(t) is a decreasing function for all t > 0, because $a'(t) = -7.2t^{-1.4} < 0$ for all t > 0.

3. $x(t) = \frac{3}{1.6}t^{1.6} + C = 5t^{1.6} + C$

   $x(0) = -50 \Rightarrow -50 = 0 + C \Rightarrow C = -50$
   $x(t) = 5t^{1.6} - 50$

4. $x(10) = 24.6450... \approx 24.6$ ft, or about three-fourths of the way through the intersection.

5. Trapezoidal rule:

   $x(10) = -50 + 2 \cdot (\frac{0}{2} + 4.5 + 6.9 + 8.8 + 10.4 + \frac{11.9}{2})$
   $= 23.1$ ft

6. Solve $100 = \frac{3}{1.6}t^{1.6} - 50$

   $t = \sqrt[1.6]{\frac{1.6}{3} \cdot 150} = 15.4678... \approx 15.5$ sec

7. [Learn]

Exploration 27: Differentials, and Linearization of a Function

1. $f(1) = \sec 1 = 1.8508...$
   $f'(1) = \sec 1 \tan 1 = 2.8824...$
   $y = 1.8508... + 2.8824...(x - 1)$

2. Graph, showing that the line is tangent to the secant curve at x = 1.

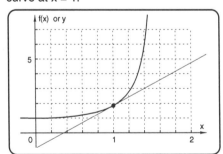

3.
x	y	f(x)	error
0.95	1.7066...	1.7191...	0.0124...
0.96	1.7355...	1.7436...	0.0081...
0.97	1.7643...	1.7689...	0.0046...
0.98	1.7931...	1.7952...	0.0020...
0.99	1.8219...	1.8225...	0.0005...
1.00	1.8508...	1.8508...	0.0
1.01	1.8796...	1.8801...	0.0005...
1.02	1.9084...	1.9107...	0.0022...
1.03	1.9372...	1.9424...	0.0051...
1.04	1.9661...	1.9754...	0.0093...
1.05	1.9949...	2.0097...	0.0148...

As x approaches 1, the error goes to zero.

4. Solving f(x) − y = 0.001 numerically, the error will be less than 0.001 if 0.9862... < x < 1.0134..., so if x is kept within 0.0134... of 1, then the error will be less than 0.01.

5. Graph, showing dx = 0.4, dy being the corresponding rise to the linear function graph, and Δy being the corresponding rise to the function graph.

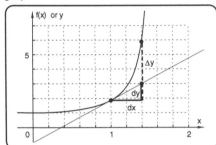

6. $dy \div dx = \frac{f'(1) \cdot (x - 1)}{(x - 1)} = f'(1)$.

   So dy ÷ dx is the same as the derivative, $\frac{dy}{dx}$.

7. dx = Δx always. dy = f'(x) dx
   or: dy is the rise along the tangent line and Δy is the rise along the graph itself.
   or: dy is close to Δy if dx is small.

8. [Learn]

Exploration 28: Riemann Sums for Definite Integrals

1. About 15.4 squares, so the area is about 308.
   (Equation is f(x) = 10 + 2x + 10 cos (0.5x), so precise answer is 308.3103....)

2.
c	f(c)
3	17
5	12
7	15
9	26
11	39
13	46

3. Graph, showing the rectangles.

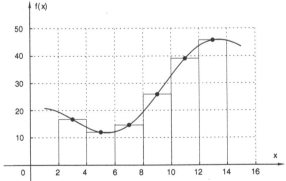

For each rectangle, the part of the graph above the rectangle is roughly the same as the part of the rectangle above the graph, so the area of each rectangle is about the same as the area under the

Calculus Explorations
© 1998 Key Curriculum Press

corresponding portion of the graph. Thus the sum of the areas of the rectangles represents the definite integral.

4. Area of rectangles
   = 2 · 17 + 2 · 12 + 2 · 15 + 2 · 26 + 2 · 39 + 2 · 46
   = 310, which is close to the 308 in Problem 1.

5.

c	f(c)
2	19
4	14
6	12
8	19
10	33
12	44

6. Area of rectangles
   = 2 · 19 + 2 · 14 + 2 · 12 + 2 · 19 + 2 · 33 + 2 · 44
   = 282

7. Both methods approximate the height of the function within each subinterval, then multiply that height by the width of the subinterval.

8. [Learn]

Exploration 29: The Mean Value Theorem

1. $f(3) = 2.3$, $f(7) = 4.3$, slope of secant line = 0.5
   $c \approx 5.5$ (numerically, graphically, or by solving $-0.3c^2 + 2.4c - 3.6 = 0.5$)
   Graph, showing tangent line parallel to secant.

   $c \approx 5.5$ (Exactly $4 + \frac{1}{3}\sqrt{21}$)

   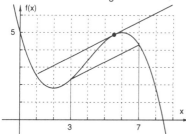

2. Graph, showing two tangents parallel to secant.

   $c \approx 2.3$ or 5.7 (Exactly $4 \pm \sqrt{3}$)

   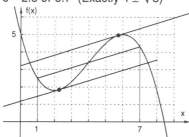

3. No such $x = c$ exists. The slope of the secant line is positive. The slope of the tangent line is greater than the slope of the secant when x is between 1 and 4, and is negative when x is between 4 and 5.

4. $g(1) = 6 - 2(-3)^{2/3}$, $g(4) = 6$
   Slope of secant line = $2(-3)^{-1/3}$
   $c = 3.1111$ (numerically, graphically, or by solving

   $2(-3)^{-1/3} = \frac{-4}{3}(c-4)^{-1/3} \Rightarrow c = 4 - \frac{8}{9}$)

Graph, showing tangent line parallel to secant.

$c \approx 3.1$ (Exactly $3\frac{1}{9}$)

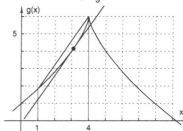

5. h is differentiable for all $x \in (5,8)$, but h is not continuous at $x = 5$ since $\lim_{x \to 5^-} h(x) \neq \lim_{x \to 5^+} h(x)$.

   $h(5) = 2.8$, $h(8) = 6.2$,

   slope of secant line = $\frac{3.4}{3} = 1.1333\ldots$

   $h'(c) = 0.4(c - 2) > 1.2 > 1.1333\ldots$ for all $c \in (5, 8)$, so no such c exists.
   Graph, showing that there is no tangent between $x = 5$ and $x = 8$ that is parallel to the secant.

   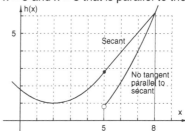

6. Graph, showing that there is a tangent line at $c \approx 3$ that parallels the secant line. ($c = 3$ exactly.)

   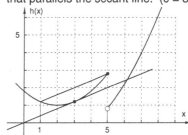

7. If $f(x)$ is (1) differentiable on (a, b), and (2) continuous at $x = a$ and $x = b$, then there is a number $x = c$ between a and b for which
   $$f'(c) = \frac{f(b) - f(a)}{b - a}.$$
   If a function satisfies the hypotheses, then that is enough (sufficient) to make the function satisfy the conclusion. But for a function that satisfies the conclusion, it is not necessary for the hypotheses to be satisfied. For example, the function in Problem 6 between $x = 2$ and $x = 7$ does not meet the hypotheses of the mvt, but these are not necessary for $c = 3$ to satisfy the conclusion.

Exploration 30: Some Very Special Riemann Sums

1. Trapezoidal rule:

   $I \approx 0.5 + \sqrt{2} + \sqrt{3} + 1 = 4.6462\ldots$
   Underestimates I because parts of the region lie above the trapezoids.

2. Midpoint sum:

   $I \approx \sqrt{1.5} + \sqrt{2.5} + \sqrt{3.5} = 4.6767\ldots$
   Overestimates I because the part of the region above the top of each rectangle is slightly less than the part of each rectangle above the region.

3. $I \approx \sqrt{1.48584256} + \sqrt{2.49161026} + \sqrt{3.49402772} = 4.6666\ldots$

   This estimate is between the Trapezoidal and Midpoint sums.

4. $I \approx \frac{1}{2}(\sqrt{1.24580513} + \sqrt{1.74701361} + \sqrt{2.24768040}$

   $+ \sqrt{2.74810345} + \sqrt{3.24839587} + \sqrt{3.74861006}) = 4.6666\ldots$, same as in Problem 3.

5. $g(1) = \frac{2}{3}$, $g(1.5) = \sqrt{1.5}$

   Slope of secant line $= \dfrac{\sqrt{1.5} - \frac{2}{3}}{0.5} = \sqrt{6} - \frac{4}{3}$

   $g'(c) = c^{1/2} = \sqrt{6} - \frac{4}{3} \Rightarrow c = \left(\sqrt{6} - \frac{4}{3}\right)^2 =$

   $1.24580513\ldots$, the sample point for $[1, 1.5]$ in Problem 4!

6. $g'(x) = f(x)$

7. Conjecture: $I = 4\frac{2}{3}$

8. [Learn]

Exploration 31: The Fundamental Theorem of Calculus

1. See the text.
2. See the text.
3. g is differentiable on (a, b) because it has a derivative f, and g is continuous on (a, b) because differentiability implies continuity.

4. There is a point $c_1 \in (a, x_1)$, such that

   $g'(c_1) = \dfrac{g(x_1) - g(a)}{\Delta x}$

   Graph, showing $x = c_1$ with tangent line parallel to secant line.

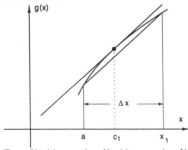

5. $R_n = f(c_1)(x_1 - a) + f(c_2)(x_2 - x_1) + f(c_3)(x_3 - x_2)$
   $+ \ldots + f(c_n)(b - x_{n-1})$
   $= (f(c_1) + f(c_2) + f(c_3) + \ldots + f(c_n))\Delta x$

6. $R_n = (g'(c_1) + g'(c_2) + g'(c_3) + \ldots + g'(c_n))\Delta x$

   $= \left(\dfrac{g(x_1) - g(a)}{\Delta x} + \dfrac{g(x_2) - g(x_1)}{\Delta x}\right.$

   $\left. + \ldots + \dfrac{g(b) - g(x_{n-1})}{\Delta x}\right)\Delta x$

   $= g(x_1) - g(a) + g(x_2) - g(x_1) + g(x_3) - g(x_2)$
   $+ \ldots + g(b) - g(x_{n-1})$

   $= g(b) - g(a)$

7. $L_n \le R_n \le U_n \Rightarrow L_n \le g(b) - g(a) \le U_n$ Problem 6

   $\lim\limits_{n\to\infty} L_n \le g(b) - g(a) \le \lim\limits_{n\to\infty} U_n$ Properties of limits
   $(g(b) - g(a)$ is a constant.)
   But the two limits equal the definite integral, by definition.

   $\therefore \int_a^b f(x)\, dx = g(b) - g(a)$, Q.E.D.

8. $g(x) = \int x^{1/2}\, dx = \frac{2}{3} x^{3/2} + C$

   $\int_1^4 x^{1/2}\, dx = \frac{2}{3} 4^{3/2} + C - \frac{2}{3} 1^{3/2} - C = 4\frac{2}{3}$
   Note that the C's cancel out!

Exploration 32: Some Properties of Definite Integrals

1. For all x, $f(-x) = f(x)$.
   [Even powers of x (e.g., x^2) have this property.]

2. $\int_{-3}^{3}(19 - x^2)\, dx = 19x - \frac{1}{3}x^3\Big|_{x=-3}^{x=3} = 96$

   $\int_0^3 (19 - x^2)\, dx = 19x - \frac{1}{3}x^3\Big|_{x=0}^{x=3} = 48$
   The region is symmetrical with respect to the y-axis. Thus the area to the left of the axis equals the area to the right, meaning that the total area is twice the area to the right.

3. For all x, $g(-x) = -g(x)$.
   [Odd powers of x (e.g., x^1) have this property.]

4. $\int_{-4}^{4} x^3\, dx = \frac{1}{4}x^4\Big|_{x=-4}^{x=4} = 0$

   The two parts of the region are congruent, and thus have the same area. Since the left part of the region is below the y-axis, the integral is negative, and thus cancels out the integral for the right part.

5. $\int_1^3 h(x)\, dx = \int_1^3 0.4x^3 - 3x^2 + 5x + 5\, dx$

   $= 0.1x^4 - x^3 + 2.5x^2 + 5x\Big|_{x=1}^{x=3}$

   $= 12$

   $\int_3^6 h(x)\, dx = \int_3^6 0.4x^3 - 3x^2 + 5x + 5\, dx$

   $= 0.1x^4 - x^3 + 2.5x^2 + 5x\Big|_{x=3}^{x=6}$

   $= 15$

6. $\int_1^6 h(x)\, dx = \int_1^3 h(x)\, dx + \int_3^6 h(x)\, dx = 12 + 15 = 27$

7. a. $\int_2^7 (u(x) + v(x))\, dx = \int_2^7 u(x)\, dx + \int_2^7 v(x)\, dx = 42$

   b. $\int_2^7 v(x)\, dx = \int_2^4 v(x)\, dx + \int_4^7 v(x)\, dx$

Calculus Explorations
© 1998 Key Curriculum Press

$$\Rightarrow 13 = 8 + \int_4^7 v(x)\, dx$$

$$\int_4^7 v(x)\, dx = 5$$

8. [Learn]

Exploration 33: Applications of Definite Integrals

1. Graph, showing sample point t on the t-axis and the corresponding point (t, v(t)) on the graph.

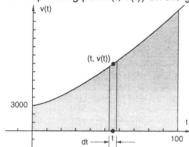

2. Graph, at Problem 1, showing strip of width dt.
3. Distance = rate · time
   If the rate throughout the time interval is essentially constant at v(t), and the length of time is dt, then
   $D = v(t) \cdot dt$

4. $R_n = \sum\limits_{i=1}^{n} v(c_i)\, dt$

5. Each term in the upper sum is greater than or equal to the corresponding $v(c_i)\, dt$ in the Riemann sum, by definition of an upper sum, so that $U_n \geq R_n$.
   Similarly, $L_n \leq R_n$.
   Thus $L_n \leq R_n \leq U_n$.
   $\therefore\ \lim\limits_{n \to \infty} L_n \leq \lim\limits_{n \to \infty} R_n \leq \lim\limits_{n \to \infty} U_n$.
   If the function is integrable, the limits of the upper and lower sums equal the definite integral, by the definition of definite integral.
   By the squeeze theorem, the limit of R_n is also equal to the definite integral. Thus
   $$\lim_{n \to \infty} R_n = \int_0^{100} v(t)\, dt, \quad \text{Q.E.D.}$$

6. $\int_0^{100} 3000 + 18t^{1.4}\, dt = 3000t + \dfrac{18}{2.4} \cdot t^{2.4}\Big|_{t=0}^{t=100}$

   $= 300{,}000 + 7.5 \cdot 100^{2.4} = 773{,}218.0083\ldots$
   ≈ 773 thousand ft

7. Solve $3000 + 18t^{1.4} = 26{,}000$
   $t = 165.5376\ldots \approx 166$ sec.

8. Dist. $\approx \int_0^{166} 3000 + 18t^{1.4}\, dt = 3000t + \dfrac{18}{2.4} \cdot t^{2.4}\Big|_{t=0}^{t=166}$

   $= 498{,}000 + 7.5 \cdot 166^{2.4} = 2{,}095{,}057.9488\ldots$
   ≈ 2.095 million ft

9. Both the f(x) and the dx have physical meanings. The dx is multiplied by f(x) to get the part of the value of the integral corresponding to the vertical strip.

Exploration 34: Derivation of Simpson's Rule

1. Graph showing $f(x) = ax^2 + bx + c$ from $x = -h$ to $x = h$.

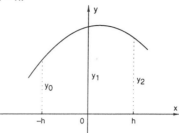

2. Area $= \int_{-h}^{h} (ax^2 + bx + c)\, dx$

   $= \dfrac{1}{3}ax^3 + \dfrac{1}{2}bx^2 + cx\Big|_{x=-h}^{x=h}$

   $= \dfrac{1}{3}ah^3 + \dfrac{1}{2}bh^2 + ch + \dfrac{1}{3}ah^3 - \dfrac{1}{2}bh^2 + ch$

   $= \dfrac{1}{3}h(2ah^2 + 6c)$

3. $y_0 = f(-h) = ah^2 - bh + c$
   $y_1 = f(0) = c$
   $y_2 = f(h) = ah^2 + bh + c$
   $y_0 + y_2 = 2ah^2 + 2c$

4. Area $= \dfrac{1}{3}h(2ah^2 + 6c)$

   $= \dfrac{1}{3}h(2ah^2 + 2c + 4c)$

   $= \dfrac{1}{3}h((y_0 + y_2) + 4y_1)$

5. Graph, $y = 50 \cdot 2^x$.

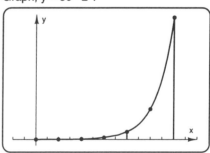

   $\int_0^{12} 50 \cdot 2^x\, dx \approx \dfrac{1}{3}h(y(0) + 4y(2) + y(4)) +$

   $\qquad \dfrac{1}{3}h(y(4) + 4y(6) + y(8)) +$

   $\qquad \dfrac{1}{3}h(y(8) + 4y(10) + y(12)) =$

   $\qquad \dfrac{1}{3}(2)(50 \cdot 2^0 + 4 \cdot 50 \cdot 2^2 + 50 \cdot 2^4) +$

   $\qquad \dfrac{1}{3}(2)(50 \cdot 2^4 + 4 \cdot 50 \cdot 2^6 + 50 \cdot 2^8) +$

   $\qquad \dfrac{1}{3}(2)(50 \cdot 2^8 + 4 \cdot 50 \cdot 2^{10} + 50 \cdot 2^{12})$

   $= 300{,}300$

6. $\int_a^b y\, dx \approx \dfrac{1}{3}h(y_0 + 4y_1 + y_2)$

   $\qquad + \dfrac{1}{3}h(y_2 + 4y_3 + y_4)$

   $\qquad + \ldots + \dfrac{1}{3}h(y_{n-2} + 4y_{n-1} + y_n)$

   $= \dfrac{1}{3}h(y_0 + 4y_1 + 2y_2 + 4y_3 + 2y_4 + \ldots + y_n)$

---

7. $\int_0^{12} 50 \cdot 2^x \, dx \approx 295{,}391.8881\ldots$ for $n = 100$

8. $T_{100} = 295{,}562.0962\ldots$
   $M_{100} = 295{,}306.6736\ldots$
   S_{100} is the closest.

9. Each parabola requires two subintervals.

The Calculus of Exponential and Logarithmic Functions

Exploration 35: Another Form of the Fundamental Theorem

1. $f(9) = \int_{t=1}^{t=9} t^{1/2} \, dt = \frac{2}{3} t^{3/2} \Big|_{t=1}^{t=9} = \frac{2}{3}(27 - 1) = \frac{52}{3}$

2. Graph, showing $f(x) = \int_{t=1}^{t=x} t^{1/2} \, dt$.

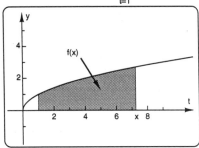

3. $f(x) = \int_{t=1}^{t=x} t^{1/2} \, dt = \frac{2}{3} t^{3/2} \Big|_{t=1}^{t=x} = \frac{2}{3}(x^{3/2} - 1)$

4. $f'(x) = \frac{d}{dx} \frac{2}{3}(x^{3/2} - 1) = x^{1/2}$

   $f'(x)$ is the integrand of $f(x)$ evaluated at x!

5. $g(x) = \int_{t=3}^{t=x} t^3 \, dt \Rightarrow g'(x) = x^3$

6. $h(x) = \int_{t=2}^{t=x^3} \cos t \, dt = \sin t \Big|_{t=2}^{t=x^3} = (\sin x^3 - \sin 2)$

7. $h'(x) = \frac{d}{dx}(\sin x^3 - \sin 2) = \cos x^3 \cdot 3x^2$
   $= 3x^2 \sin x^3$

8. The pattern is: The integrand evaluated at the upper limit of integration, times derivative of the upper limit function (the "inside" function), which is an example of the chain rule.
   $\frac{d}{dx}\left(\int_{t=0}^{t=\sin x} \tan^3 (t^5) \, dt\right) = \tan^3 (\sin^5 x) \cdot \cos x$

9. In the statement of the ftc in the problem, $g(x) = L(x)$, and $f(x) = 1/t$, so $L'(x) = 1/x$.

10. The power rule for integration works only with exponents not equal to -1.

Exploration 36: Natural Logs and the Uniqueness Theorem

1. $\ln x = \int_{t=1}^{t=x} 1/t \, dt$

2. $f(7) = \ln 21 = \int_{t=1}^{t=21} 1/t \, dt = 3.0445\ldots$ (numerically)

$g(7) = \ln 7 + \ln 3 = \int_{t=1}^{t=7} 1/t \, dt + \int_{t=1}^{t=3} 1/t \, dt$
   $= 1.9459\ldots + 1.0986\ldots$ (numerically)
   $= 3.0445\ldots = f(7)!$

3. $f(1) = \ln 3$
   $g(1) = \ln 1 + \ln 3$
   but $\ln 1 = \int_{t=1}^{t=1} 1/t \, dt = 0$, so $f(1) = g(1)$.

4. $f'(x) = \frac{1}{3x} \cdot 3 = \frac{1}{x}$
   $g'(x) = 0 + \frac{1}{x} = \frac{1}{x}$
   $\therefore$ f and g are differentiable on $(1, b)$ for any $b > 0$

5. f and g are differentiable for all $x > 0$.
   Thus they are differentiable at $x = 1$ and $x = b$.
   Since differentiability implies continuity, f and g are continuous at $x = a$ and $x = b$.

6. The sum or difference of continuous functions is continuous, and the sum or difference of differentiable functions is differentiable.

7. $h(1) = f(1) - g(1) = 0$.
   The mvt says that for some c between 1 and b,
   $h'(c) = \frac{h(b) - h(1)}{b - 1} = \frac{h(b)}{b - 1}$.

8. $h(b) \neq 0$, so $h'(c) \neq 0$.

9. $h'(x) = f'(x) - g'(x) = \frac{1}{x} - \frac{1}{x} = 0$ for all $x > 0$.
   But then $h'(c) = 0$ since c is between 1 and b. Therefore the assumption made in Problem 8 must have been false, and there is no such number $b > 0$ for which $f(b) \neq g(b)$.
   Therefore, $f(x) = g(x)$ for all $x > 0$.

10. See the text.

11. [Learn]

Exploration 37: Properties of Logarithms

1. $f(x) = \ln 7x$, $g(x) = \ln 7 + \ln x$
   $f'(x) = \frac{1}{7x} \cdot 7 = \frac{1}{x}$ and $g'(x) = 0 + \frac{1}{x} = \frac{1}{x}$
   $\therefore$ $f'(x) = g'(x)$ for all $x > 0$.
   $f(1) = \ln 7$ and $g(1) = \ln 7 + \ln 0 = \ln 7$
   $\therefore$ $f(1) = g(1)$.
   $\therefore$ $f(x) = g(x)$ for all $x > 0$, Q.E.D.

2. $p(x) = \ln 7 - \ln x$ and $q(x) = \ln \frac{7}{x}$
   $p'(x) = 0 - \frac{1}{x} = -\frac{1}{x}$ and $q'(x) = \frac{x}{7} \cdot \frac{1}{7}(-x^{-2}) = -\frac{1}{x}$
   $\therefore$ $p'(x) = q'(x)$ for all $x > 0$.
   $p(1) = \ln 7 - \ln 1 = \ln 7$ and $q(1) = \ln 7$
   $\therefore$ $p(1) = q(1)$
   $\therefore$ $p(x) = q(x)$ for all $x > 0$, Q.E.D.

Calculus Explorations
© 1998 Key Curriculum Press

3. $h(x) = \ln(x^5)$ and $j(x) = 5 \ln x$

$h'(x) = \dfrac{1}{x^5} \cdot 5x^4 = \dfrac{5}{x}$ and $j'(x) = 5 \cdot \dfrac{1}{x} = \dfrac{5}{x}$

$\therefore h'(x) = j'(x)$ for all $x > 0$.

$h(1) = \ln 1 = 0$ and $j(1) = 5 \ln 1 = 5 \cdot 0 = 0$

$\therefore h(1) = j(1)$.

$\therefore h(x) = j(x)$ for all $x > 0$, Q.E.D.

4. $r = \log_s t \Leftrightarrow s^r = t$

5. $m^p = w \Leftrightarrow p = \log_m w$

6. Let $y = \log_b x$.

$b^y = x$

$\log_a b^y = \log_a x$

$y \log_a b = \log_a x$

$y = \dfrac{\log_a x}{\log_a b} \Rightarrow \log_b x = \dfrac{\log_a x}{\log_a b}$, Q.E.D.

7. $\ln x = \log_e x = \dfrac{\log_{10} x}{\log_{10} e}$

Then $\log_{10} e = \dfrac{\log_{10} x}{\ln x}$.

Let $x = 2$, for example.

$\log_{10} e = \dfrac{\log_{10} 2}{\ln 2} = 0.43429\ldots$

$e = 10^{0.43429\ldots} = 2.718281828\ldots$

8. [Learn]

Exploration 38: Derivative of an Exponential Function

1. $y = 5^x$

Numerical derivative at $x = 2$ is $40.2359\ldots$.

2. The power rule for 5^x is $x \cdot 5^{x-1}$.

At $x = 2$, $2 \cdot 5^{x-1} = 10$, which does not equal $40.2359\ldots$.

3. $\ln y = \ln 5^x = x \ln 5$

$\dfrac{1}{y} y' = \ln 5$

$y' = y \ln 5 = 5^x \ln 5$

4. $y'(2) = 5^2 \ln 5 = 40.2359\ldots$ Correct answer!

5. $f(x) = b^x \Rightarrow f'(x) = b^x \ln b$

6. $f(x) = 5(0.6^x)$

$f'(x) = 5(0.6^x) \ln 0.6$

$f'(1) = 5(0.6) \ln 0.6 = -1.5324\ldots$

Graph, showing that the line through $(1, f(1))$ with slope $-1.5324\ldots$ is tangent to the graph.

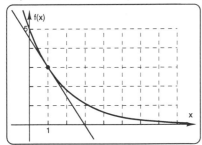

7. $\displaystyle\int 5^x \, dx = \dfrac{1}{\ln 5} \cdot 5^x + C$

8. [Learn]

Exploration 39: Base e Logs vs. Natural Logs

1. Graph, $y = (1 + 1/x)^x$.

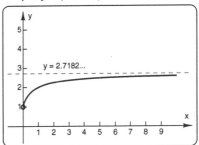

2. $y = 2.71813082\ldots$ at $x = 9000$, but $y = 2.71814592\ldots$ at $x = 10000$. The line is not horizontal.

3. $e^1 = 2.71828182\ldots$

Conjecture: $\displaystyle\lim_{x \to \infty} (1 + 1/x)^x = e$

4. If $b = e$, then $\log_b e = 1$

Thus $f'(x) = \dfrac{d}{dx} \log_b x = \dfrac{1}{x}$

5. $f(x) = \log_e x$ and $g(x) = \ln x$

$f'(x) = 1/x$ and $g'(x) = 1/x$.

$\therefore f'(x) = g'(x)$ for all $x > 0$.

$f(1) = 0$ and $g(1) = 0$.

$\therefore f(1) = g(1)$.

$\therefore f(x) = g(x)$ for all $x > 0$, Q.E.D.

6. $x = \log_e y \Rightarrow y = e^x$

7. If $f(x) = \ln x$, then $f^{-1}(x) = e^x$

8. [Learn]

Exploration 40: A Compound Interest Problem

1. $10{,}000 = M(0) = ae^{k0} = a \cdot 1$, so $a = 10{,}000$.

$15{,}528.08 = M(4) = ae^{k4} = 10{,}000e^{k4}$

$\Rightarrow 1.552808 = e^{4k} \Rightarrow 4k = \ln 1.552808$

$\Rightarrow k = 0.11001622\ldots$

2. $M'(0) = \$1100.1622\ldots$ / year (numerically)

$M'(4) = \$1708.3407\ldots$ / year (numerically)

3. $\dfrac{M'(0)}{M(0)} = \dfrac{\$1100.1622\ldots}{\$10000.00} = 0.11001622\ldots$

$\dfrac{M'(4)}{M(4)} = \dfrac{\$1708.3407\ldots}{\$15528.08} = 0.11001622\ldots$

Both answers are equal to k!

4. $\dfrac{d}{dt}(M(t)) = ae^{kt} \cdot k$

$\therefore \dfrac{\frac{d}{dt}(M(t))}{M(t)} = \dfrac{ae^{kt} \cdot k}{ae^{kt}} = k$

This is the exponential constant, the coefficient of x in the exponent!

5. $M(8) = 10{,}000e^{8k} = \$24{,}112.13$

6. $M(50) = 10{,}000e^{50k} = \$2{,}448{,}905.34$

7. Solve $20{,}000 = 10{,}000e^{kt}$ for t:

$2 = e^{kt} \Rightarrow kt = \ln 2 \Rightarrow t = \dfrac{\ln 2}{k} = 6.3004\ldots$ years.

8. [Learn] (Among other things, start saving early for your financial goals!)

Exploration 41: A Limit by L'Hospital's Rule

1. $x = 2 \cos t - \cos 2t$
   $y = 2 \sin t + \sin 2t$

   At $t = \dfrac{\pi}{3}$,

   $x = 2 \cos \dfrac{\pi}{3} - \cos \dfrac{2\pi}{3} = 2 \cdot 0.5 - (-0.5) = 1.5$

   $y = 2 \sin \dfrac{\pi}{3} + \sin \dfrac{2\pi}{3} = 2 \cdot \dfrac{\sqrt{3}}{2} + \dfrac{\sqrt{3}}{2} = \dfrac{3}{2} \cdot \sqrt{3} = $
   $2.5980\ldots$

   These are the coordinates of the cusp shown in the graph.

2. $\dfrac{dx}{dt} = -2 \sin t + 2 \sin 2t$

   $\dfrac{dy}{dt} = 2 \cos t + 2 \cos 2t$

   $\dfrac{dy}{dx} = \dfrac{dy/dt}{dx/dt} = \dfrac{2 \cos t + 2 \cos 2t}{-2 \sin t + 2 \sin 2t}$ (chain rule)

   $\phantom{\dfrac{dy}{dx}} = \dfrac{\cos t + \cos 2t}{-\sin t + \sin 2t}$

3. $\dfrac{dx}{dt}\left(\dfrac{\pi}{3}\right) = -2 \sin \dfrac{\pi}{3} + 2 \sin \dfrac{2\pi}{3} = -2 \cdot \dfrac{\sqrt{3}}{2} + 2 \cdot \dfrac{\sqrt{3}}{2} = 0$

   $\dfrac{dy}{dt}\left(\dfrac{\pi}{3}\right) = 2 \cos \dfrac{\pi}{3} + 2 \cos \dfrac{2\pi}{3} = 2 \cdot 0.5 + 2(-0.5) = 0$

   $\dfrac{dy}{dx}$ takes the indeterminate form $\dfrac{0}{0}$ using the parametric chain rule.

4.

t	dy/dx
$\pi/3 - 0.005$	$1.7420\ldots$
$\pi/3 - 0.004$	$1.7400\ldots$
$\pi/3 - 0.003$	$1.7380\ldots$
$\pi/3 - 0.002$	$1.7360\ldots$
$\pi/3 - 0.001$	$1.7340\ldots$
$\pi/3$	
$\pi/3 + 0.001$	$1.7300\ldots$
$\pi/3 + 0.002$	$1.7280\ldots$
$\pi/3 + 0.003$	$1.7260\ldots$
$\pi/3 + 0.004$	$1.7240\ldots$
$\pi/3 + 0.005$	$1.7220\ldots$

   $\dfrac{dy}{dx}$ seems to approach $1.7320\ldots$ as $x \to \dfrac{\pi}{3}$.

5. Graph, showing line with slope 1.7320 through the cusp. The line is tangent to each branch of the graph.

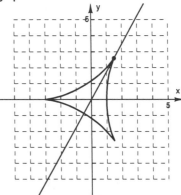

6. $\displaystyle\lim_{t\to\pi/3} \dfrac{\cos t + \cos 2t}{-\sin t + \sin 2t}$

   $= \displaystyle\lim_{t\to\pi/3} \dfrac{\dfrac{d}{dt}(\cos t + \cos 2t)}{\dfrac{d}{dt}(-\sin t + \sin 2t)}$

   $= \displaystyle\lim_{t\to\pi/3} \dfrac{-\sin t - 2 \sin 2t}{-\cos t + 2 \cos 2t}$

   $= \dfrac{-\sin \pi/3 - 2 \sin 2\pi/3}{-\cos \pi/3 + 2 \cos 2\pi/3}$

   $= \dfrac{-\sqrt{3/2} - \sqrt{3}}{-0.5 - 1.0} = \sqrt{3} = 1.7320\ldots$

   Same as Problem 4.

7. $\displaystyle\lim_{x\to 0} 1 - 5x - \cos 3x = 1 - 0 - \cos 0 = 0 = \displaystyle\lim_{x\to 0} x$, so l'Hospital's rule applies.

   $\displaystyle\lim_{x\to 0} \dfrac{1 - 5x - \cos 3x}{x}$

   $= \displaystyle\lim_{x\to 0} \dfrac{\dfrac{d}{dx}(1 - 5x - \cos 3x)}{\dfrac{d}{dx}(x)}$

   $= \displaystyle\lim_{x\to 0} \dfrac{0 - 5 + 3 \sin 3x}{1}$

   $= -5$

8. [Learn]

Calculus Explorations
© 1998 Key Curriculum Press

The Calculus of Growth and Decay

Exploration 42: Differential Equation for Compound Interest

1. Let M be the amount of money at time t.
   $\frac{dM}{dt} = kM$, where k is a constant.

2. $\frac{dM}{dt} = kM \Rightarrow M^{-1}\, dM = k\, dt$
   $\int M^{-1}\, dM = \int k\, dt$
   $\ln|M| = kt + C$
   $|M| = e^{kt+C}$
   $M = \pm e^{kt+C}$

3. $M = \pm e^{kt} \cdot e^{C}$

4. A positive number raised to any power is positive. Thus e^{C} is positive.

5. If C_1 can be positive or negative, then when $M > 0$ let $C_1 = e^{C}$, and when $M < 0$ (say, for debt outstanding on a credit card) let $C_1 = -e^{C}$.

6. $M(0) = C_1 e^{k \cdot 0} = C_1 \Rightarrow C_1 = 1000$

7. $k = 0.05$

8. $M(5) = 1000e^{0.25} \approx \$1284.03$
   $M(10) = 1000e^{0.5} \approx \$1648.72$
   $M(50) = 1000e^{2.5} \approx \$12{,}182.49$
   $M(100) = 1000e^{5} \approx \$148{,}413.16$

9. $2000 = 1000e^{0.05t} \Rightarrow t = \frac{\ln 2}{0.05} = 13.8629\ldots$ years

10. [Learn]

Exploration 43: Differential Equation for Memory Retention

1. $\frac{dy}{dt} = R - ky$

2. $(R - ky)^{-1}\, dy = dt$

3. $\int (R - ky)^{-1}\, dy = \int dt$
   $-\frac{1}{k} \int (R - ky)^{-1} (-k\, dy) = \int dt$
   $\frac{-1}{k} \cdot \ln|R - ky| = t + C$

4. $\ln|R - ky| = -kt + C_1$
   $|R - ky| = e^{-kt + C_1} = e^{C_1} \cdot e^{-kt}$
   $R - ky = \pm e^{C_1} \cdot e^{-kt} = C_2\, e^{-kt}$
   $ky = R - C_2\, e^{-kt}$

5. $k \cdot 0 = R - Ce^{-k \cdot 0} \Rightarrow C = R$
   $\therefore\ ky = R - Re^{-kt}$
   $y = \frac{R}{k} \cdot (1 - e^{-kt})$

6. $R = 100;\ ky = 4$ when $y = 10 \Rightarrow k = 0.4$
   $y = 250(1 - e^{-0.4t})$

7. $y(3) = 250(1 - e^{-1.2}) = 174.7014\ldots \approx 175$ names

8. [Learn]

Exploration 44: Introduction to Slope Fields

1. $\frac{dy}{dx}(5,2) = -0.9,\ \frac{dy}{dx}(-8,9) = 0.32$
   Graph showing that the slopes at these points look reasonably close to −0.9 and 0.32.

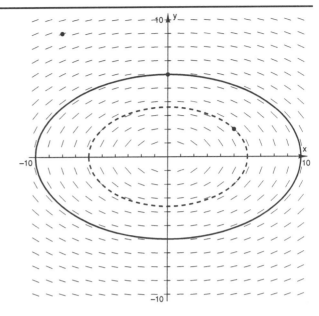

2. Graph, solid line at Problem 1, showing that below the x-axis the graph curves back on itself and closes. The figure appears to be an ellipse.

3. Graph, dashed line at Problem 1. The figure appears to be another ellipse inside the first one.

4. $\frac{dy}{dx} = -\frac{0.36x}{y} \Rightarrow y\, dy = -0.36x\, dx$
   $\int y\, dy = \int -0.36x\, dx$
   $\frac{1}{2}y^2 = -\frac{1}{2} \cdot 0.36x^2 + C$
   $\left(\frac{x}{10}\right)^2 + \left(\frac{y}{6}\right)^2 = C_1$
   This is a standard form of the equation of an ellipse centered at the origin with x- and y-radii equal to $10\sqrt{C_1}$ and $6\sqrt{C_1}$, respectively.

5. [Learn]

Exploration 45: Euler's Method

1. $\frac{dy}{dx} = -\frac{x}{2y}$
   $\frac{dy}{dx}(0,3) = 0$
   $dx = 0.5$ and $\frac{dy}{dx} = 0 \Rightarrow$
   $dy = \frac{dy}{dx} \cdot dx = 0(0.5) = 0$
   $y \approx 3 + dy = 3$ at $x = 0 + dx = 0.5$

2. $\frac{dy}{dx}(0.5,3) = -\frac{1}{12}$
   $dy = -\frac{1}{12}\, dx = -\frac{1}{24}$
   $y \approx 3 + dy = 2\frac{23}{24}$ at $x = 0.5 + dx = 1$

3.

x	y	slope	dy
0	3	0	0
0.5	3	−0.0833...	−0.0416...
1	2.9583...	−0.1690...	−0.0845...
1.5	2.8738...	−0.2609...	−0.1304...
2	2.7433...	−0.3645...	−0.1822...
2.5	2.5610...	−0.4880...	−0.2440...
3	2.3170...	−0.6473...	−0.3236...
3.5	1.9933...	−0.8779...	−0.4389...
4	1.5543...	−1.2866...	−0.6433...
4.5	0.9110...	−2.4696...	−1.2348...
5	−0.3237...	7.7213...	3.8606...
5.5	3.5368...	−0.7775...	−0.3887...
6	3.1481...	−0.9529...	−0.4764...
6.5	2.6716...	−1.2164...	−0.6082...
7	2.0634...	−1.6962...	−0.8481...

(It is recommended that students write or download a program to do Euler's method as soon as they have seen the pattern for calculating the y-values. That way they can concentrate on implications such as the discontinuity beyond x = 4.5, not on the tedious computations.)

Sample EULER Program for TI-82

```
Disp "FIRST X"        Y+Y₁D→Y
Input X               X+D→X
Disp "FIRST Y"        Disp "X"
Input Y               Disp X
Disp "DELTA X"        Disp "Y"
Input D               Disp Y
Lbl 1                 Pause
                      Goto 1
```

4. Graph (through dots), showing that Euler's method seems to produce reasonable answers for x < 5, where the slope first becomes positive.

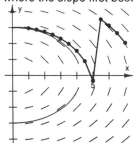

5. $\dfrac{dy}{dx} = -\dfrac{x}{2y} \Rightarrow 2y\,dy = -x\,dx \Rightarrow y^2 = -\dfrac{1}{2}x^2 + C$

$3^2 = -\dfrac{1}{2}0^2 + C \Rightarrow C = 9$

$\dfrac{1}{2}x^2 + y^2 = 9$

Graph, at Problem 4. The figure is an ellipse with x-radius $\sqrt{18} = 4.2426...$ and y-radius 3. When the curve approaches the x-axis, the slope becomes nearly infinite. Furthermore, after crossing the x-axis the curve should go in the negative x-direction, but the procedure does not account for this.

6. [Learn]

Exploration 46: A Predator-Prey Problem

1. If there are relatively few coyotes, then one would expect the population of deer to increase. The graph will start by going to the right. Consequently, it also goes up, in order to follow the slope lines. Note: The differential equation used is

$$\dfrac{dy}{dx} = \dfrac{-0.64(x - 65)}{y - 35} .$$

2. Graph, containing (80, 20). Both populations will increase until the population of coyotes gets too high. Then the deer population will begin to decrease, eventually causing the coyotes to starve off. The coyote population will then decrease to where the deer population can begin growing again, and the situation will return to where it started. The behavior is "cyclical."

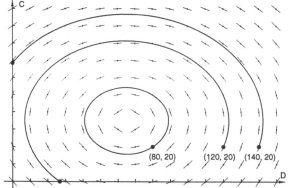

3. Graph at Problem 2, containing (120, 20). The graph eventually intersects the D-axis, at which point the coyotes become extinct. There are so many deer to start with that too many coyotes are born, and the deer are hunted so intensely that the surviving deer population is too small to support any coyotes, which die off completely. Assuming the coyotes are their only natural enemy, the deer population will grow in an unrestrained way! (The graph, however, goes along the D-axis since there are no coyotes.)

4. Graph at Problem 2, containing (140, 20). The graph eventually intersects the C-axis and the deer become extinct. There are so many deer to start with that way too many coyotes are born, and all the deer are eaten. Assuming that the deer are the coyotes' only food source, the coyote population will eventually die off by starvation. (The graph would go down the C-axis to the origin.)

5. At (100,35), the curve has vertical tangent, so that the deer born just balance the deer killed; but the population of coyotes is increasing.
   At (65,80), the curve has horizontal tangent, so that the coyotes born just balance the coyotes dying. The population of deer is decreasing.

6. At (65,35) neither population is increasing or decreasing.

7. [Learn]

Calculus Explorations
© 1998 Key Curriculum Press

Exploration 47: Maxima, Minima, and Points of Inflection

1. $y = 5x^{2/3} - x^{5/3}$

   $y' = \frac{10}{3}x^{-1/3} - \frac{5}{3}x^{2/3}$

   $= \frac{5}{3}x^{-1/3}(2 - x)$

2. y' is zero only when $x = 2$. At this x, the tangent line is horizontal, the graph stops increasing and starts decreasing. As shown on the given figure, the graph has a maximum at $x = 2$.

3. y' is infinite at $x = 0$ because $0^{-1/3}$ is equivalent to $1/0^{1/3}$, which involves division by zero. The slope becomes infinite as x approaches zero from either direction. As shown on the given graph, there is a cusp at $x = 0$, where the value of y is a local minimum.

4. $y' = \frac{10}{3}x^{-1/3} - \frac{5}{3}x^{2/3}$

   $y'' = -\frac{10}{9}x^{-4/3} - \frac{10}{9}x^{-1/3}$

   $= -\frac{10}{9}x^{-4/3}(1 + x)$

5. At $x = 1$, $y'' = \frac{-20}{9} < 0$.

   Graphically, this means that the "hollowed out" side of the graph faces downward.

6. y'' is zero only when $x = -1$.

7. $y''(-2) = -\frac{10}{9} \cdot \sqrt[3]{16} \cdot -1 = \frac{20}{3}\sqrt[3]{2} > 0$

   Thus the graph is concave up at $x = -2$.

   $y''(-0.5) = -\frac{10}{9} \cdot \sqrt[3]{0.0625} \cdot 0.5 = -\frac{5}{18}\sqrt[3]{0.5} < 0$

   Thus the graph is concave down at $x = -0.5$.

8. [Learn]

Exploration 48: Derivatives and Integrals from Given Graphs

1. Graph, $h'(x)$.
   (Equation is $h(x) = x + 4$ for $x \leq 1$,
   $h(x) = 0.2x^2 - 1.6x + 6.4$ for $x \geq 1$.) Graph accounts for $h'(x)$ undefined at $x = 1$ by the open circles on the ends of the branches.

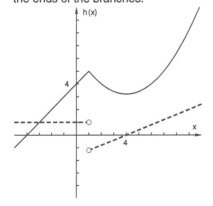

2. $h'(x)$ is not infinite at $x = 1$. The graph stays bounded on a neighborhood of 1.

3. h is continuous, but not differentiable, at $x = 1$.

4. Graph, $g(x)$.
   (Equation is $g'(x) = 0.5x^2 - x - 1.5$)

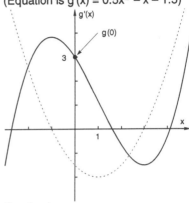

5. Graph, y'.
   (Equation is $y = x$ for $x \leq 1$, $y = 1$ for $1 \leq x \leq 2$, and $y = \sqrt{1 - (x - 3)^2}$ for $2 \leq x \leq 4$ (a semicircle).)

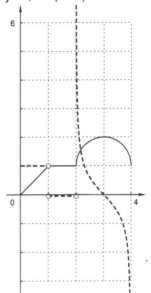

6. Graph, $\int y'\,dx$. Note that the graph is concave up for $2 < x < 3$ and concave down for $3 < x < 4$. (Same equation as in Problem 5)

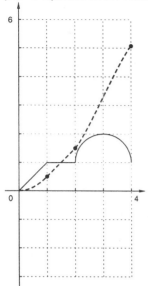

7. [Learn]

Exploration 49: Maximal Cylinder in a Cone Problem

1. $V = \pi r^2 h = \pi x^2 y$
   At $x = 0$, $y = 12$, $V = 0$ in^3
   At $x = 1$, $y = 9$, $V = 9\pi$ in^3
   At $x = 2$, $y = 6$, $V = 24\pi$ in^3
   At $x = 3$, $y = 3$, $V = 27\pi$ in^3
   At $x = 4$, $y = 0$, $V = 0$ in^3

2. For the point (x, y) shown in the diagrams,
   $y = 12 - 3x$.
   So, the radius of the inscribed cylinder is x, the altitude is $y = 12 - 3x$, and the volume is
   $V = \pi r^2 h = \pi x^2 (12 - 3x) = 12\pi x^2 - 3\pi x^3$.

3. $\frac{dV}{dx} = 24\pi x - 9\pi x^2$

4. $0 = 24\pi x - 9\pi x^2 = 9\pi x\left(\frac{24}{9} - x\right)$
   $\frac{dV}{dx} = 0$ at $x = 0$ and $x = \frac{24}{9} = \frac{8}{3}$.

5. $V = 12\pi x^2 - 3\pi x^3$:

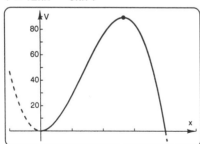

   Maximum volume occurs at $x = \frac{8}{3}$. The other solution, $x = 0$, is a local minimum.

6. At $x = \frac{8}{3}$, $r = \frac{8}{3}$, $h = 4$, $V = \frac{256}{9}\pi = 28.4444\ldots\pi$ in^3.

---

7. • Find an equation for the quantity to be maximized.
   • Get the equation in terms of one independent variable.
   • Find the derivative.
   • Find critical points where the derivative is zero (or undefined).
   • Find out whether the critical points correspond to maximum or to minimum values.
   • Answer the question that was asked.

8. [Learn]

Exploration 50: Volume by Plane Disk Slices

1. The volume of a disk is $\pi r^2 h$, where r = radius and h = thickness (think of a disk as a very short cylinder). The slice shown is not exactly a disk, since the radius changes ever so slightly from the top to the bottom. But the radius of this slice is always very close to x, and the height is dx, so the volume is approximately $dV = \pi x^2\,dy$. (In the limit, the difference between the cylinders and the actual disk volumes approaches zero.)

2. $y = 4 - x^2 \Rightarrow x^2 = 4 - y$, so
   $dV = \pi x^2\,dy = \pi(4 - y)\,dy$.

3. $V = \int_{y=0}^{y=4} \pi \cdot (4 - y)\,dy$
   $= 4\pi y - \frac{1}{2}\pi y^2 \Big|_{y=0}^{y=4}$
   $= 8\pi$

4. The circumscribed cylinder has $r = 2$, $h = 4$, and $V_{cyl} = \pi r^2 h = 16\pi$.
   The inscribed cone has $r = 2$, $h = 4$, and $V_{cone} = \frac{1}{3}\pi r^2 h = 5\frac{1}{3} \cdot \pi$.
   The calculated volume of the paraboloid falls between these numbers.

5. Conjecture: The volume of a paraboloid is half the volume of the circumscribed cylinder.

6. [Learn]

Exploration 51: Volume by Plane Washer Slices

1. $A_{washer} = A_{outer\ circle} - A_{inner\ circle}$
   $A = \pi y_1^2 - \pi y_2^2 = \pi(y_1^2 - y_2^2)$
   $dV = A\,dx = \pi(y_1^2 - y_2^2)\,dx$

2. $y_1 = 6e^{-0.2x}$, $y_2 = \sqrt{x}$
   $dV = \pi(36e^{-0.4x} - x)\,dx$

3. $V = \int_{x=1}^{x=4} \pi(36e^{-0.4x} - x)\,dx$
   $= -90\pi e^{-0.4x} - \frac{1}{2}\pi x^2 \Big|_{x=1}^{x=4}$
   $= -90\pi e^{-1.6} + 90\pi e^{-0.4} - 7.5\pi$

4. $V = 108.8816\ldots$

5. Numerical integration gives $108.8816\ldots$, the same answer!

6. [Learn]

Exploration 52: Volume by Cylindrical Shells

1. The circumference is $2\pi x$, so
   $$dV = C \cdot h \cdot dx = 2\pi x \cdot y \cdot dx$$
2. $y = 4x - x^2$, so
   $$dV = 2\pi x \cdot (4x - x^2)\, dx = 8\pi x^2 - 2\pi x^3.$$
3. $V = \int_{x=0}^{x=3} (8\pi x^2 - 2\pi x^3)\, dx$

   $= \frac{8}{3}\pi x^3 - \frac{1}{2}\pi x^4 \Big|_{x=0}^{x=3}$

   $= 31.5\pi$

   As x goes from 0 to 3, the shells generate the whole figure. The curve shown for negative values of x is just the *image* of the graph as it rotates around, not the graph itself.
4. Slice the region parallel to the x-axis.
   Pick a sample point (x, y) on the graph, within the slice.
   As the region rotates, the slice generates a cylindrical shell with radius y, altitude x, and thickness dy. Thus,
   $$dV = 2\pi y \cdot x \cdot dy$$
   Since $y = \ln x$, $x = e^y$. So
   $$dV = 2\pi y e^y\, dy$$
   $$V = \int_{y=1}^{y=2} 2\pi y e^y\, dy$$
   $= 46.4268\ldots$ (numerically).
   (Exact answer is $2\pi e^2$, integrating by parts.)
5. [Learn]

Exploration 53: Length of a Plane Curve (Arc Length)

1. Graph, showing three line segments approximating the graph.

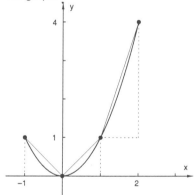

   Dist. $(-1, 1)$ to $(0, 0) = \sqrt{1^2 + 1^2} = \sqrt{2}$

   Dist. $(0, 0)$ to $(1, 1) = \sqrt{1^2 + 1^2} = \sqrt{2}$

   Dist. $(1, 1)$ to $(2, 4) = \sqrt{1^2 + 3^2} = \sqrt{10}$

   Length of curve is about $2\sqrt{2} + \sqrt{10} = 5.9907\ldots$
   This method underestimates the length of the graph, because it fails to account for the curvature of the graph between the sample points.
2. Dist. $(-1, 1)$ to $(-0.5, 0.25)$

   $= \sqrt{0.5^2 + 0.75^2} = \sqrt{0.8125}$

   Dist. $(-0.5, 0.25)$ to $(0, 0)$

   $= \sqrt{0.5^2 + 0.25^2} = \sqrt{0.3125}$

   Dist. $(0, 0)$ to $(0.5, 0.25)$

   $= \sqrt{0.5^2 + 0.25^2} = \sqrt{0.3125}$

   Dist. $(0.5, 0.25)$ to $(1, 1)$

   $= \sqrt{0.5^2 + 0.75^2} = \sqrt{0.8125}$

   Dist. $(1, 1)$ to $(1.5, 2.25)$

   $= \sqrt{0.5^2 + 1.25^2} = \sqrt{1.8125}$

   Dist. $(1.5, 2.25)$ to $(2, 4)$

   $= \sqrt{0.5^2 + 1.75^2} = \sqrt{3.3125}$

   Length of curve is about

   $\sqrt{0.3125} + \sqrt{0.8125} + \sqrt{1.8125} + \sqrt{3.3125}$
   $= 6.0871\ldots$
   This estimate is better because the many shorter line segments better approximate the curve than the few long line segments.
3. The estimate gets better as Δx gets smaller, so to get the exact length, take the limit as $\Delta x \to 0$.
4. $\Delta L = \sqrt{\Delta x^2 + \Delta y^2} = \sqrt{1 + [\Delta y/\Delta x]^2}\, \Delta x$, but the mean value theorem states that for some c in the sample interval, $f'(c) = \Delta y/\Delta x$.
5. Since $\Delta x = dx$ and $\Delta y \approx dy = f'(x)\, dx$, dL can be written $dL = \sqrt{1 + (f'(x))^2}\, dx = \sqrt{dx^2 + (f'(x))^2 dx^2}$
   $= \sqrt{dx^2 + dy^2}$, where dy is evaluated at the sample point $x = c$.
6. $y = x^2 \Rightarrow dy = 2x\, dx$
   $dL = \sqrt{dx^2 + dy^2} = \sqrt{dx^2 + (2x\, dx)^2}$
   $= \sqrt{1 + 4x^2}\, dx$
   $L = \int_{x=-1}^{x=2} dL = \int_{x=-1}^{x=2} \sqrt{1 + 4x^2}\, dx$
7. $L = 6.1257\ldots$ (numerically)
   $(\sqrt{17} + \frac{1}{4}\ln(\sqrt{17} + 4) + \frac{1}{2}\sqrt{5} - \frac{1}{4}\ln(\sqrt{5} - 2)$ is exact.)
8. [Learn]

Exploration 54: Area of a Surface of Revolution

1. Graph, showing a narrow slice whose surface resembles a frustum of a cone.

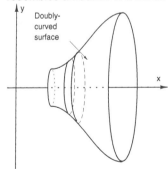

Doubly-curved surface

2. The arc length of the base of the cone is $2\pi R$, so if you slit the cone down one side, and "unroll" it, the resulting section of a circle will have radius L and arc length $2\pi R$. A full circle of radius L would have arc length $2\pi L$, and area πL^2.
   But this section is only $\frac{2\pi R}{2\pi L}$ of a full circle, so it has area $\frac{2\pi R}{2\pi L} \cdot \pi L^2 = \pi R L$, Q.E.D.

3. First note that $\frac{r}{l} = \frac{R}{L}$ because of similar triangles, so $Rl - rL = 0$. Then using problem 2, the frustum has area:

$A = \pi RL - \pi rl$
$= \pi RL - \pi Rl + \pi Rl - \pi rL + \pi rL - \pi rl$
$= \pi RL - \pi Rl + \pi(0) + \pi rL - \pi rl$
$= \pi R(L - l) + \pi r(L - l)$
$= \pi(R + r)(L - l)$

or: $A = \pi RL - \pi rl = \pi R(L - \frac{r}{R} \cdot l) = \pi R(L - \frac{l}{L} \cdot l)$

$= \pi\frac{R}{L}(L^2 - l^2) = \pi\frac{R}{L}(L + l)(L - l)$

$= \pi(R + R \cdot \frac{l}{L})(L - l) = \pi(R + R \cdot \frac{r}{R})(L - l)$

$= \pi(R + r)(L - l)$

But the average radius is $\frac{1}{2}(R + r)$, the circumference at this radius is $\pi(R + r)$, and the slant height of the frustum is $(L - l)$, Q.E.D.

4. The slice is approximately a frustum of slant height dL and average radius y. So the surface area is about $2\pi y\, dL$.

5. $y = x^3 \Rightarrow dy = 3x^2\, dx$

$dL = \sqrt{dx^2 + dy^2} = \sqrt{1 + 9x^4}\, dx$

$dS = 2\pi y\, dL = 2\pi x^3\sqrt{1 + 9x^4}\, dx$

$S = \int_0^1 2\pi x^3\sqrt{1 + 9x^4}\, dx$

$= \frac{1}{18}\pi \int_0^1 (1 + 9x^4)^{1/2}\,(36x^3\, dx)$

$= \frac{1}{18}\pi \cdot \frac{2}{3}(1 + 9x^4)^{3/2}\,\big|_0^1$

$= \frac{\pi}{27}(10^{1.5} - 1)$

$= 3.5631\ldots$

6. [Learn]

Exploration 55: Area of an Ellipse in Polar Coordinates

1. Graph, which agrees with the given figure.

2. $r(0.3) = \dfrac{10}{3 - 2\cos 0.3} = 9.1799\ldots$

   $x(0.3) = r(0.3) \cdot \cos 0.3 = 8.7699\ldots$
   $y(0.3) = r(0.3) \cdot \sin 0.3 = 2.71286\ldots$
   The sample point shown is at about $(8.8, 2.7)$. The angle measures about $17°$, which is $0.296\ldots$ radian. So all three values agree with the graph.

3. At $\theta = 0.3$, the radius (of the sector of circle which the wedge approximates) is $r(0.3)$, and the central angle is $d\theta = 0.1$, so the approximate area is $\frac{0.1}{2\pi} \cdot \pi \cdot 9.1799\ldots^2 = 0.4589\ldots$ units.

4. The sector is approximately $\frac{dq}{2\pi}$ of a circle of radius r and area πr^2, so $dA = \frac{dq}{2\pi} \cdot \pi r^2 = \frac{1}{2}r^2\, d\theta$.

5. $A = \displaystyle\int_{\theta = 0}^{\theta = 2\pi} \frac{1}{2}\left(\frac{10}{3 - 2\cos\theta}\right)^2 d\theta$
   $= 84.2977\ldots$ (numerically)

6. The y-radius is approximately 4.5, so the area formula predicts that the area will be $\pi \cdot 6 \cdot 4.5 = 84.8230\ldots$. Close!

   (y-radius is exactly $6\sin(\cos^{-1}\frac{2}{3}) = 4.4721\ldots$, making the area $\pi \cdot 6 \cdot 4.4721\ldots = 84.2977\ldots$, which is precisely the numerical answer.)

7. [Learn]

Algebraic Calculus Techniques for the Elementary Functions

Exploration 56: Integration by Parts Practice

1. $\int x^2 \sin 3x\, dx$

	u	dv
	x^2	$\sin 3x$
	$2x$	$-\frac{1}{3}\cos 3x$
	2	$-\frac{1}{9}\sin 3x$
	0	$\frac{1}{27}\cos 3x$

$= -\frac{1}{3}x^2\cos 3x + \frac{2}{9}x\sin 3x + \frac{2}{27}\cos 3x + C$

2. $\int x^5 \ln 4x\, dx$

	u	dv
	$\ln 4x$	x^5
	x^{-1}	$\frac{1}{6}x^6$
	1	$\frac{1}{6}x^5$
	0	$\frac{1}{36}x^6$

$= \frac{1}{6}x^6 \ln 4x - \frac{1}{36}x^6 + C$

3. $\int e^{5x}\cos 6x\, dx$

	u	dv
	e^{5x}	$\cos 6x$
	$5e^{5x}$	$\frac{1}{6}\sin 6x$
	$25e^{5x}$	$-\frac{1}{36}\cos 6x$

$= \frac{1}{6}e^{5x}\sin 6x + \frac{5}{36}e^{5x}\cos 6x - \frac{25}{36}\int e^{5x}\cos 6x\, dx$

$\frac{61}{36}\int e^{5x}\cos 6x\, dx = \frac{1}{6}e^{5x}\sin 6x + \frac{5}{36}e^{5x}\cos 6x + C$

$\int e^{5x}\cos 6x\, dx = \frac{6}{61}e^{5x}\sin 6x + \frac{5}{61}e^{5x}\cos 6x + C_1$

4. $\int x\,(\ln x)^3\, dx$

	u	dv
	$(\ln x)^3$	x
	$3(\ln x)^2\, x^{-1}$	$\frac{1}{2}x^2$
	$3(\ln x)^2$	$\frac{1}{2}x$
	$6(\ln x)\, x^{-1}$	$\frac{1}{4}x^2$
	$6\ln x$	$\frac{1}{4}x$
	$6x^{-1}$	$\frac{1}{8}x^2$
	6	$\frac{1}{8}x$
	0	$\frac{1}{16}x^2$

$= \frac{1}{2}x^2 (\ln x)^3 - \frac{3}{4}x^2 (\ln x)^2 + \frac{3}{4}x^2 \ln x - \frac{3}{8}x^2 + C$

5. $\int \sin^{10} x\,(\cos x\, dx) = \frac{1}{11}\sin^{11} x + C$

6. $\int \sin^{10} x\, dx$

	u	dv
	$\sin^9 x$	$\sin x$
	$-9\sin^8 x\cos x$	$-\cos x$

$= -\sin^9 x\cos x + 9\int \sin^8 x\cos^2 x\, dx$
$= -\sin^9 x\cos x + 9\int \sin^8 x\,(1 - \sin^2 x)\, dx$
$= -\sin^9 x\cos x + 9\int \sin^8 x\, dx - 9\int \sin^{10} x\, dx$
$10\int \sin^{10} x\, dx = -\sin^9 x\cos x + 9\int \sin^8 x\, dx$
$\int \sin^{10} x\, dx = -\frac{1}{10}\sin^9 x\cos x + \frac{9}{10}\int \sin^8 x\, dx$

7. [Learn]

Exploration 57: Reduction Formulas

1. $\int \sin^{10} x \, dx$

$$\begin{array}{ccc} & u & dv \\ & \sin^9 x & \sin x \\ & -9\sin^8 x \cos x & -\cos x \end{array}$$

$= -\sin^9 x \cos x + 9 \int \sin^8 x \cos^2 x \, dx$
$= -\sin^9 x \cos x + 9 \int \sin^8 x \, (1 - \sin^2 x) \, dx$
$= -\sin^9 x \cos x + 9 \int \sin^8 x \, dx - 9 \int \sin^{10} x \, dx$
$10 \int \sin^{10} x \, dx = -\sin^9 x \cos x + 9 \int \sin^8 x \, dx$
$\int \sin^{10} x \, dx = -\frac{1}{10} \sin^9 x \cos x + \frac{9}{10} \int \sin^8 x \, dx$

2. $\int \sin^n x \, dx = -\frac{1}{n} \sin^{n-1} x \cos x + \frac{n-1}{n} \int \sin^{n-2} x \, dx$
   $(n \neq 0)$

3. $V = \int_0^\pi \pi \, (\sin^3 x)^2 \, dx$

$= \pi \int_0^\pi \sin^6 x \, dx$
$= 3.0842\ldots$ (numerically)

4. $V = \int_0^\pi \pi \, (\sin^3 x)^2 \, dx$

$= \pi \int_0^\pi \sin^6 x \, dx$

$\int \sin^6 x \, dx = -\frac{1}{6} \sin^5 x \cos x + \frac{5}{6} \int \sin^4 x \, dx$

$= -\frac{1}{6} \sin^5 x \cos x$
$\qquad + \frac{5}{6}(-\frac{1}{4} \sin^3 x \cos x + \frac{3}{4} \int \sin^2 x \, dx)$
$= -\frac{1}{6} \sin^5 x \cos x - \frac{5}{24} \sin^3 x \cos x$
$\qquad + \frac{15}{24}(-\frac{1}{2} \sin x \cos x + \frac{1}{2}x) + C$
$= -\frac{1}{6} \sin^5 x \cos x - \frac{5}{24} \sin^3 x \cos x$
$\qquad\qquad - \frac{5}{16} \sin x \cos x + \frac{5}{16}x + C$

$V = \pi(-\frac{1}{6} \sin^5 x \cos x - \frac{5}{24} \sin^3 x \cos x$
$\qquad\qquad - \frac{5}{16} \sin x \cos x + \frac{5}{16}x) \Big|_0^\pi$
$= \pi(0 - 0 - 0 + \frac{5}{16}\pi + 0 + 0 + 0 + 0)$
$= \frac{5}{16}\pi^2 = 3.0842\ldots$

5. [Learn]

Exploration 58: Integrals of Special Powers of Trigonometric Functions

1. $\int \cos^7 x \, (\sin x \, dx) = \frac{1}{8} \cos^8 x + C$
2. The differential $\sin x \, dx$ of $\cos x$ does not appear in the integrand.
3. $\int \cos^7 x \, dx = \int (\cos^2 x)^3 \cos x \, dx$

$= \int (1 - \sin^2 x)^3 \cos x \, dx$
$= \int (1 - 3\sin^2 x + 3\sin^4 x - \sin^6 x) \cos x \, dx$
$= \int \cos x \, dx - 3 \int \sin^2 x \cos x \, dx$
$\qquad + 3 \int \sin^4 x \cos x \, dx - \int \sin^6 x \cos x \, dx$
$= \sin x - \sin^3 x + \frac{3}{5} \sin^5 x - \frac{1}{7} \sin^7 x + C$

4. The power of $\cos x$ is even, so converting to powers of $(1 - \sin^2 x)$ leaves no $\cos x \, dx$ to be the differential of $\sin x$.
5. $\int \tan^5 x \, (\sec^2 x \, dx) = \frac{1}{6}\tan^6 x + C$
6. The differential $\sec x \tan x \, dx$ of $\sec x$ does not appear in the integrand.
7. $\int \sec^8 x \, dx = \int (\sec^2 x)^3 \sec^2 x \, dx$

$= \int (\tan^2 x + 1)^3 \sec^2 x \, dx$
$= \int (\tan^6 x + 3\tan^4 x + 3\tan^2 x + 1) \sec^2 x \, dx$
$= \int \tan^6 x \sec^2 x \, dx + 3 \int \tan^4 x \sec^2 x \, dx$
$\qquad + 3 \int \tan^2 x \sec^2 x \, dx + \int \sec^2 x \, dx$
$= \frac{1}{7}\tan^7 x + \frac{3}{5}\tan^5 x + \tan^3 x + \tan x + C$

8. The power of $\sec x$ is odd, so converting to powers of $(\tan^2 x + 1)$ does not leave $\sec^2 x \, dx$ to be the differential of $\tan x$.
9. [Learn]

Exploration 59: Other Special Trigonometric Integrals

1. $\cos 2x = \cos (x + x) = \cos^2 x - \sin^2 x$
2. $\cos 2x = \cos^2 x - \sin^2 x$
   $= \cos^2 x - (1 - \cos^2 x)$
   $= 2\cos^2 x - 1$
3. $\cos 2x = \cos^2 x - \sin^2 x$
   $= (1 - \sin^2 x) - \sin^2 x$
   $= 1 - 2\sin^2 x$
4. $\cos 2x = 2\cos^2 x - 1$
   $\cos^2 x = \frac{1}{2}(1 + \cos 2x)$
   $\cos 2x = 1 - 2\sin^2 x$
   $\sin^2 x = \frac{1}{2}(1 - \cos 2x)$
5. $\int \cos^2 x \, dx = \int (\frac{1}{2} + \frac{1}{2}\cos 2x) \, dx$

$= \frac{1}{2}x + \frac{1}{4}\sin 2x + C$
$\int \sin^2 x \, dx = \int (\frac{1}{2} - \frac{1}{2}\cos 2x) \, dx$
$= \frac{1}{2}x - \frac{1}{4}\sin 2x + C$

6. $\int \cos 7x \sin 5x \, dx$

$$\begin{array}{ccc} & u & dv \\ & \cos 7x & \sin 5x \\ & -7\sin 7x & -\frac{1}{5}\cos 5x \\ & -49\cos 7x & -\frac{1}{25}\sin 5x \end{array}$$

$= -\frac{1}{5} \cos 7x \cos 5x - \frac{7}{25} \sin 7x \sin 5x$
$\qquad\qquad + \frac{49}{25} \int \cos 7x \sin 5x \, dx$
$-\frac{24}{25} \int \cos 7x \sin 5x \, dx$
$\qquad = -\frac{1}{5} \cos 7x \cos 5x - \frac{7}{25} \sin 7x \sin 5x + C$
$\int \cos 7x \sin 5x \, dx$
$\qquad = \frac{5}{24} \cos 7x \cos 5x + \frac{7}{24} \sin 7x \sin 5x + C_1$

7. $\int \cos ax \sin bx \, dx$

$= \frac{a}{a^2-b^2} \sin ax \sin bx + \frac{b}{a^2-b^2} \cos ax \cos bx + C$

8. [Learn]

Exploration 60: Introduction to Integration by Trigonometric Substitution

1. $A = 2 \int_{-2}^{7} \sqrt{64 - x^2}\, dx = 126.9622\ldots$ (numerically)

2. The differential $2x\, dx$ of the "inside function" $64 - x^2$ does not appear in the integrand, so the integral cannot be done as the antiderivative of a power.

3. $\frac{x}{8} = \sin \theta \Rightarrow x = 8 \sin \theta$

   $dx = 8 \cos \theta\, d\theta$
   By the Pythagorean theorem, the horizontal leg of the right triangle is $\sqrt{64 - x^2}$.
   Thus $8 \cos \theta = \sqrt{64 - x^2}$.
   $\therefore \int \sqrt{64 - x^2}\, dx = \int (8 \cos \theta) \cdot 8 \cos \theta\, d\theta$
   $= 64 \int \cos^2 \theta\, d\theta$

4. $64 \int \cos^2 \theta\, d\theta = 32 \int (1 + \cos 2\theta)\, d\theta$
   $= 32\theta + 16 \sin 2\theta + C$

5. $64 \int \cos^2 \theta\, d\theta = 32\theta + 16(2 \sin \theta \cos \theta) + C$

   $= 32 \sin^{-1} \frac{x}{8} + 32 \cdot \frac{x}{8} \cdot \frac{\sqrt{64 - x^2}}{8} + C$

   $= 32 \sin^{-1} \frac{x}{8} + \frac{1}{2}x \sqrt{64 - x^2} + C$

6. $A = 2 \int_{-2}^{7} \sqrt{64 - x^2}\, dx = 64 \sin^{-1} \frac{x}{8} + x\sqrt{64 - x^2}\, \Big|_{-2}^{7}$

   $= 64 \sin^{-1} \frac{7}{8} + 7\sqrt{15} - 64 \sin^{-1} \frac{-1}{4} + 2\sqrt{60}$

   $= 126.9622\ldots,$
   which agrees with the answer found numerically in Problem 1.

7. [Learn]

Exploration 61: Integrals of Rational Functions by Partial Fractions

1. $\frac{A}{x - 5} + \frac{B}{x + 1} = \frac{10x - 32}{(x - 5)(x + 1)}$

   $\Rightarrow A(x + 1) + B(x - 5) = 10x - 32$
   $\Rightarrow A + B = 10,\ A - 5B = -32$
   $\Rightarrow A = 3,\ B = 7$

2. $\int \frac{10x - 32}{(x - 5)(x + 1)}\, dx = \int \frac{3}{x - 5}\, dx + \int \frac{7}{x + 1}\, dx$

   $= 3 \ln |x - 5| + 7 \ln |x + 1| + C$

3. (Instructor input on Heaviside method.)

4. See your text for reasons behind the Heaviside method.

5. $x^3 - 2x^2 - 5x + 6 = (x - 1)(x + 2)(x - 3)$

   $\int \frac{11x^2 - 22x - 13}{(x - 1)(x + 2)(x - 3)}\, dx$

   $= \int \frac{4}{x - 1}\, dx + \int \frac{5}{x + 2}\, dx + \int \frac{2}{x - 3}\, dx$

   $= 4 \ln |x - 1| + 5 \ln |x + 2| + 2 \ln |x - 3| + C$

6. [Learn]

Exploration 62: Chain Experiment

1. (The following is simulated data, agreeing with the figure shown on the Exploration. Actual data will depend on the chain used.)
   Vertex: (0, 20)
   Left end: (−90, 120), Right end: (90, 120)

2. $y = k \cosh \frac{1}{k}x + C$

   $20 = k \cosh 0 + C \Rightarrow 20 = k + C$
   $120 = k \cosh \frac{90}{k} + C$
   Substitute $20 - k$ for C.
   $120 = k \cosh \frac{90}{k} + 20 - k$
   $0 = k \cosh \frac{90}{k} - 100 - k$
   Solving numerically gives $k = 51.7801\ldots$
   $C = 20 - 51.7801\ldots = -31.7801\ldots$
   Equation is:

   $y = 51.7801\ldots \cosh \frac{1}{51.7801\ldots}x - 31.7801\ldots$

3. A typical medium-weight tow chain weighs about 0.015 lb/cm, which is the value of w.
   $h = (51.7801\ldots)(0.015) \approx 0.8$ pounds

4.
x	y
0	20
20	23.9
40	36.2
60	58.8
80	95.1

5. Graph, showing that the chain fits closely the data points.

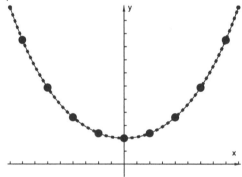

6. $dL = \sqrt{dx^2 + dy^2}$

   $= \sqrt{1 + \sinh^2 \frac{1}{k}x}\ dx$

   $L = \int_{-90}^{90} dL = 285.3490\ldots \approx 285$ cm
   The measured length should be close to this.

7. See the text derivation.

8. [Learn]

Exploration 63: Introduction to Improper Integrals

1. For Calvin, $v(t) = 320(t + 4)^{-1}$, so distance d(t) is
   $d(t) = \int 320(t + 4)^{-1} dt = 320 \ln (t + 4) + C$ for $t \geq 0$
   $d(0) = 0 \Rightarrow C = -320 \ln 4$
   $d(t) = 320(\ln (t + 4) - \ln 4)$
   $d(10) = 320(\ln 14 - \ln 4) = 400.8841\ldots \approx 401$ ft
   $d(20) = 320(\ln 24 - \ln 4) = 573.3630\ldots \approx 573$ ft
   $d(50) = 320(\ln 54 - \ln 4) = 832.8606\ldots \approx 833$ ft

Calculus Explorations
© 1998 Key Curriculum Press

2. $d(b) = 320(\ln (b + 4) - \ln 4)$ for $b \geq 0$
   $1000 = 320(\ln (b + 4) - \ln 4)$
   $b = 87.0395\ldots \approx 87$ seconds
   (Exactly $4(e^{3.125} - 1)$)

3. For Phoebe, $v(t) = 80e^{-0.1t}$, so distance $d(t)$ is
   $d(t) = \int 80e^{-0.1t}\, dt = -800e^{-0.1t} + C$.
   $d(0) = 0 \Rightarrow C = 800$
   $d(t) = -800e^{-0.1t} + 800 = 800(1 - e^{-0.1t})$
   $d(10) = 800(1 - e^{-1}) = 505.6964\ldots \approx 506$ ft
   $d(20) = 800(1 - e^{-2}) = 691.7317\ldots \approx 692$ ft
   $d(50) = 800(1 - e^{-3}) = 794.6096\ldots \approx 795$ ft

4. $d(b) = 800(1 - e^{-0.1b})$
   $1000 = 800(1 - e^{-0.1b})$
   $1.25 = 1 - e^{-0.1b}$
   $e^{-0.1b} = -0.25$, which is impossible.
   Phoebe never reaches 1000 ft!

5. $\lim\limits_{b\to\infty} 800(1 - e^{-0.1b}) = 800 - 800 \lim\limits_{b\to\infty} e^{-0.1b} = 800$
   $\lim\limits_{b\to\infty} 320(\ln (b + 4) - \ln 4) = \infty$, since $\ln x$ is unbounded.

6. It is a seeming paradox that the velocity remains positive but the distance approaches a limit! The Greeks, including Zeno of Elea (ca. 490-430 B.C.) wrestled with such paradoxes. It was the invention of calculus that allowed people to find out what happens in such situations.

7. [Learn]

Exploration 64: Miscellaneous Integration Practice!

1. $\int \tan^5 4x\, dx = \int (\tan^3 4x)(\sec^2 4x - 1)\, dx$
   $= \frac{1}{16} \tan^4 4x - \int (\tan^3 4x)\, dx$
   $= \frac{1}{16} \tan^4 4x - \int (\tan 4x)(\sec^2 4x - 1)\, dx$
   $= \frac{1}{16} \tan^4 4x - \frac{1}{8} \tan^2 4x + \int \tan 4x\, dx$
   $= \frac{1}{16} \tan^4 4x - \frac{1}{8} \tan^2 4x + \frac{1}{4} \ln |\sec 4x| + C$

2. $\int \sqrt{1 + t^2}\, dt$
   Let $t/1 = \tan \theta$.
   Thus $dt = \sec \theta\, d\theta$ and $\sqrt{1 + t^2} = \sec \theta$.

   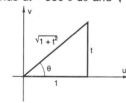

   $\int \sqrt{1 + t^2}\, dt$
   $= \int \sec \theta \sec^2 \theta\, d\theta = \int \sec^3 \theta\, d\theta$
   $= \frac{1}{2} \sec \theta \tan \theta + \frac{1}{2} \ln |\sec \theta + \tan \theta| + C$
   $= \frac{1}{2} t \sqrt{1 + t^2} + \frac{1}{2} \ln |\sqrt{1 + t^2} + t| + C$

3. $\int \tanh x\, dx = \ln (\cosh x) + C$

4. $\int x^3 e^{-x}\, dx$

u		dv
x^3	+	e^{-x}
$3x^2$	−	$-e^{-x}$
$6x$	+	e^{-x}
6	−	$-e^{-x}$
0	+	e^{-x}

   $= -e^{-x}(x^3 + 3x^2 + 6x + 6) + C$
   $\int_1^\infty x^3 e^{-x}\, dx = \lim\limits_{b\to\infty} \int_1^b x^3 e^{-x}\, dx$
   $= \lim\limits_{b\to\infty} (-e^{-b}(b^3 + 3b^2 + 6b + 6) + 6)$

   $\lim\limits_{b\to\infty} e^{-b} b^3 = \lim\limits_{b\to\infty} \frac{b^3}{e^b} \to \frac{\infty}{\infty}$
   $= \lim\limits_{b\to\infty} \frac{3b^2}{e^b} \to \frac{\infty}{\infty}$
   $= \lim\limits_{b\to\infty} \frac{6b}{e^b} \to \frac{\infty}{\infty}$
   $= \lim\limits_{b\to\infty} \frac{6}{e^b} \to \frac{6}{\infty}$
   $= 0$

   Similarly, each power of b multiplied by e^{-b} approaches zero as b approaches infinity.
   $\therefore \int_1^\infty x^3 e^{-x}\, dx = 0 + 0 + 0 + 0 + 6 = 6$

5. $\int \dfrac{x}{(x - 2)(x - 3)(x - 4)}\, dx$
   $= \int \frac{1}{x - 2}\, dx + \int \frac{-3}{x - 3}\, dx + \int \frac{2}{x - 4}\, dx$
   $= \ln |x - 2| - 3 \ln |x - 3| + 2 \ln |x - 4| + C$

6. $\int \sin^5 x\, dx = \int (1 - \cos^2 x)^2 \sin x\, dx$
   $= \int (1 - 2 \cos^2 x + \cos^4 x) \sin x\, dx$
   $= -\cos x + \frac{2}{3} \cos^3 x - \frac{1}{5} \cos^5 x + C$

7. $\int (x^4 + 2)^3\, dx = \int (x^{12} + 6x^8 + 12x^4 + 8)\, dx$
   $= \frac{1}{13} x^{13} + \frac{2}{3} x^9 + \frac{12}{5} x^5 + 8x + C$

8. $\int x^2 e^{x^3}\, dx = \frac{1}{3} \int e^{x^3} (3x^2\, dx) = \frac{1}{3} e^{x^3} + C$

9. $\int e^{ax} \cos bx\, dx$

u		dv
e^{ax}	+	$\cos bx$
ae^{ax}	−	$\frac{1}{b} \sin bx$
$a^2 e^{ax}$	+	$-\frac{1}{b^2} \cos bx$

   $= \frac{1}{b} e^{ax} \sin bx + \frac{a}{b^2} e^{ax} \cos bx$
   $\qquad\qquad - \frac{a^2}{b^2} \int e^{ax} \cos bx\, dx$
   $\frac{a^2 + b^2}{b^2} \int e^{ax} \cos bx\, dx$
   $= \frac{1}{b} e^{ax} \sin bx + \frac{a}{b^2} e^{ax} \cos bx + C$
   $\int e^{ax} \cos bx\, dx$
   $= \frac{b}{a^2 + b^2} e^{ax} \sin bx + \frac{a}{a^2 + b^2} e^{ax} \cos bx + C_1$

10. $\int \sin^{-1} ax\, dx = x \sin^{-1} ax - \int \dfrac{ax}{\sqrt{1 - (ax)^2}}\, dx$
    $= x \sin^{-1} ax + \frac{1}{a} \sqrt{1 - (ax)^2} + C \qquad (a \neq 0)$
    $\int \sin^{-1} ax\, dx = 0 + C \qquad (a = 0)$

11. [Learn]

Exploration 65: Finding Distance from Acceleration Data

1. Assume the average acceleration for the first time interval is

   $a \approx \frac{1}{2}(5 + 12) = 8.5.$

   If the initial velocity is $v = 3$, then the velocity at the end of the first time interval is

   $v = 3 + 8.5(10) = 88$ m/sec at $t = 10$.

2.
t	a	v
0	5	3
10	12	88
20	11	203
30	−4	238
40	−13	153
50	−20	−12
60	0	−112

3. Assume the average velocity for the first time interval is

   $v = \frac{1}{2}(3 + 88) = 45.5.$

   If the initial displacement is $d = 0$, then the displacement at the end of the first time interval is

   $d = 0 + 45.5(10) = 455$ m at $t = 10$.

4.
t	a	v	d
0	5	3	0
10	12	88	455
20	11	203	1910
30	−4	238	4115
40	−13	153	6070
50	−20	−12	6775
60	0	−112	6155

5. The fact that v changes sign somewhere between $t = 0$ and $t = 60$ shows that the object stops and begins going backward.

6. The object did not go back beyond its starting point since the displacement at $t = 60$ is 6155 m, which is still positive.

7. [Learn]

Exploration 66: Average Velocity

1. $v(1) = 6 − 0.3 = 5.7$

   $v(17) = 6 \cdot 17 − 0.3 \cdot 289 = 15.3$

   Both (1, 5.7) and (17, 15.3) appear to lie on the graph.

2. Displacement $= \int_{1}^{17} (6t − 0.3t^2)\, dt = 3t^2 − 0.1x^3 \Big|_{1}^{17}$

   $= 372.8$ ft.

3. Average velocity $= \dfrac{372.8 \text{ ft}}{17 − 1 \text{ sec}} = 23.3$ ft/sec.

4. No. $\frac{1}{2}(v(1) + v(17)) = 10.5$, not 23.3.

5. Graph, showing the region whose area represents Rhoda's displacement.

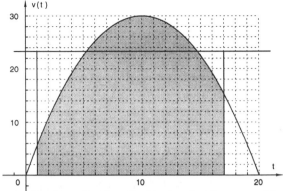

6. Graph, at Problem 5, showing that the area of the rectangle equals the area of the shaded region.

7. The shaded area above the rectangle equals the unshaded region within the rectangle.

8. There are about 26 squares under the curve for $0 \le t \le 10$, so the displacement is about 26 units, and the average velocity is $\dfrac{\text{total disp.}}{\text{time}} = \dfrac{26}{10} = 2.6.$

   (The equation is $v(t) = 6e^{-0.2t}$, so the area is 25.9399. . . , and the average is 2.5939. . . .)

9. [Learn]

Exploration 67: Introduction to Related Rates

1. Know: $\dfrac{dx}{dt} = 40$ ft/sec, $\dfrac{dy}{dt} = −30$ ft/sec

   $\dfrac{dy}{dt} < 0$ because Calvin's distance from the intersection is decreasing.

2. Want: $\dfrac{dz}{dt}$

3. $x^2 + y^2 = z^2$

4. $2x\dfrac{dx}{dt} + 2y\dfrac{dy}{dt} = 2z\dfrac{dz}{dt}$

   $\dfrac{dz}{dt} = \dfrac{x\dfrac{dx}{dt} + y\dfrac{dy}{dt}}{z}$

5. When $x = 200$ ft and $y = 600$ ft, then

   $z = \sqrt{200^2 + 600^2} = 200\sqrt{10}$ ft

   $\therefore \dfrac{dz}{dt} = \dfrac{200 \cdot 40 + 600(−30)}{200\sqrt{10}} = −5\sqrt{10}$

   $= −15.8113. . . .$

   Distance is decreasing at about 15.8 ft/sec

6. The calculations contradict the conclusion. The distance between them is actually decreasing.

7. Write down a list of known rates.

   Write a list of wanted rates.

   Write an equation relating the variables in the known and wanted rates.

   Differentiate this equation implicitly.

   Solve the resulting differential equation for the wanted rate.

   Substitute the known and calculate the values.

   Answer the question.

8. [Learn]

Exploration 68: Introduction to Minimal Path Problems

1. Length of road through clearing = $3000 - x$
   Cost of road through clearing = $20 \cdot (3000 - x)$

   Length of road through woods = $\sqrt{1000^2 + x^2}$

   Cost of road through woods = $30 \cdot \sqrt{1000^2 + x^2}$

   Total cost = $C(x) = 60,000 - 20x + 30\sqrt{1000^2 + x^2}$

2. Graph of $C(x)$.

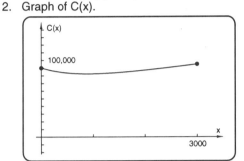

3. Cost appears to be a minimum around $x = 900$.
4. $C'(x) = -20 + 30x \cdot (1000^2 + x^2)^{-1/2}$
   $0 = C'(x) = -20 + 30x \cdot (1000^2 + x^2)^{-1/2}$

   $\Rightarrow 1.5x = \sqrt{1000^2 + x^2}$
   $\Rightarrow 2.25x^2 = 1000^2 + x^2$

   $\Rightarrow x = \dfrac{1000}{\sqrt{1.25}} = 894.4271\ldots \approx 894$ ft

   At this point the derivative goes from negative to positive, so graph stops decreasing and starts increasing.

5. Minimum cost of the road is:

   $C(1000/\sqrt{1.25}) = 60,000 + 10,000\sqrt{5}$
   $= \$82,360.68$.
   To build to the closest point on the edge of the woods:
   $C(0) = \$90,000.00$, \$7639.32 more (about 9.3% more).
   To build directly from the gate to the cabin:

   $C(3000) = 30,000\sqrt{10} = \$94,868.33$, \$12,507.65 more (about 15.2% more).

6. [Learn]

Exploration 69: Introduction to Calculus of Vectors

1. $\dfrac{dx}{dt} = -4 \sin 0.8t$, $\dfrac{dx}{dt}(1) = -4 \sin 0.8 = -2.8694\ldots$
   $\dfrac{dy}{dt} = 2.4 \cos 0.8t$, $\dfrac{dy}{dt}(1) = 2.4 \cos 0.8 = 1.6720\ldots$

2. Graph, showing dx/dt and dy/dt as dotted vectors originating on the path of the pendulum.

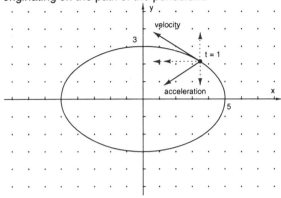

3. Graph, at Problem 2.
   The vector sum, $-2.8694\ldots\vec{i} + 1.6720\ldots\vec{j}$, is tangent to the graph at $t = 1$, and points in the direction of motion of the pendulum.

4. Speed = $|\vec{v}(1)| = \sqrt{2.8694\ldots^2 + 1.6720\ldots^2} = 3.3210\ldots$. Assuming that x and y are in feet, the velocity is about 3.32 feet per second.

5. $\dfrac{d^2x}{dt^2} = -3.2 \cos 0.8t$, $\dfrac{d^2x}{dt^2}(1) = -3.2 \cos 0.8 = -2.2294\ldots$
   $\dfrac{d^2y}{dt^2} = -1.92 \sin 0.8t$, $\dfrac{d^2y}{dt^2}(1) = -1.92 \sin 0.8 = -1.3773\ldots$
   Graph, at Problem 2, showing the second-derivative vectors in the x- and y-directions, and the resultant acceleration vector.

6. $\vec{a}(1) = -2.2294\ldots\vec{i} - 1.3773\ldots\vec{j}$
   The angle between $\vec{a}(1)$ and $\vec{v}(1)$ is smaller than 90°, so the acceleration has a component in the same direction of the velocity vector. Thus the speed is increasing.

7. Speed = $\sqrt{(dx/dt)^2 + (dy/dt)^2}$

   $\dfrac{d\text{Speed}}{dt} = \dfrac{\dfrac{d}{dt}\left((dx/dt)^2 + (dy/dt)^2\right)}{2\sqrt{(dx/dt)^2 + (dy/dt)^2}}$

   $= \dfrac{\dfrac{dx}{dt} \cdot \dfrac{d^2x}{dt^2} + \dfrac{dy}{dt} \cdot \dfrac{d^2y}{dt^2}}{\sqrt{(dx/dt)^2 + (dy/dt)^2}}$

   $\dfrac{d\text{Speed}}{dt}(1)$

   $= \dfrac{-2.8694\ldots \cdot -2.2294\ldots + 1.6720\ldots \cdot -1.3773\ldots}{3.3210\ldots}$

   $= 1.2328\ldots$ (ft/sec)/sec

8. [Learn]

Exploration 70: Derivatives of a Position Vector

1. At t = 0.5: $5 \sin 0.5 = 2.3971\ldots$, $5 \cos^2 0.5 = 3.8507\ldots$
   At t = 1: $5 \sin 1 = 4.2073\ldots$, $5 \cos^2 1 = 1.4596\ldots$
   Both points $(2.3971\ldots, 3.8507\ldots)$ and $(4.2073\ldots, 1.4596\ldots)$ are on the graph.

2. Graph, showing the position vectors at t = 0.5 and at t = 1.

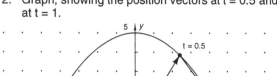

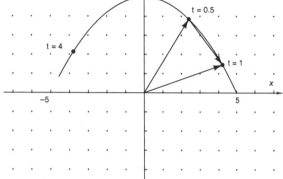

3. $\Delta \vec{r} = 1.8102\ldots \vec{i} - 2.3911\ldots \vec{j}$
   Graph, at Problem 2, showing that the head of $\Delta \vec{r}$ is at the head of $\vec{r}(1)$.

4. $\vec{v}_{av}[0.5,1]$
   $= \dfrac{5(\sin 1 - \sin 0.5)\vec{i} + 5(\cos^2 1 - \cos^2 0.5)\vec{j}}{0.5}$
   $= 3.6204\ldots \vec{i} - 4.8722\ldots \vec{j}$
   $\vec{v}_{av}[0.5,0.6]$
   $= \dfrac{5(\sin 0.6 - \sin 0.5)\vec{i} + 5(\cos^2 0.6 - \cos^2 0.5)\vec{j}}{0.1}$
   $= 4.2608\ldots \vec{i} - 4.4486\ldots \vec{j}$
   Graph, showing the average velocity vectors from t = 0.5 to t = 1 and from t = 0.5 to t = 0.6.

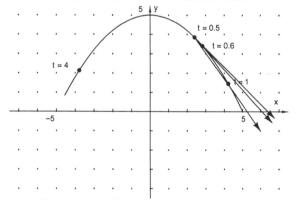

5. $\vec{v}(t) = 5 \cos t \, \vec{i} - 10 \cos t \sin t \, \vec{j} = 5 \cos t \, \vec{i} - 5 \sin 2t \, \vec{j}$
   $\vec{v}(0.5) = 5 \cos 0.5 \, \vec{i} - 5 \sin 1 \, \vec{j}$
   $= 4.3879\ldots \vec{i} - 4.2073\ldots \vec{j}$
   Graph, at Problem 4, showing $\vec{v}(0.5)$. The average velocity vectors seem to approach $\vec{v}(0.5)$ as a limit.

6. Speed $= |\vec{v}(0.5)| = 5\sqrt{\cos^2 0.5 + \sin^2 1}$
   $= 6.0791\ldots \approx 6.1\ldots$ ft/sec

7. $\vec{v}(t) = 5 \cos t \, \vec{i} - 5 \sin 2t \, \vec{j}$
   $\vec{a}(t) = -5 \sin t \, \vec{i} - 10 \cos 2t \, \vec{j}$

8. $\vec{r}(4) = -3.7840\ldots \vec{i} + 2.1362\ldots \vec{j}$
   $\vec{v}(4) = -3.2682\ldots \vec{i} - 4.9467\ldots \vec{j}$
   $\vec{a}(4) = 3.7840\ldots \vec{i} + 1.4550\ldots \vec{j}$
   Graph, showing the velocity and acceleration vectors.

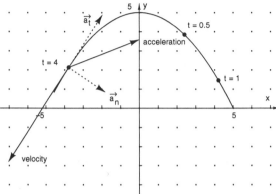

9. Speed $= |\vec{v}(4)| = 5\sqrt{\cos^2 4 + \sin^2 8} = 5.9289\ldots$
   ≈ 5.9 ft/sec

10. $\vec{a}_t(4) = \dfrac{\vec{v}(4) \cdot \vec{a}(4)}{|\vec{v}(4)|} \dfrac{1}{|\vec{v}(4)|} \vec{v}(4)$
    $= \dfrac{-25 \sin 4 \cos 4 + 50 \sin 8 \cos 8}{25(\cos^2 4 + \sin^2 8)} \cdot$
    $(5 \cos 4 \, \vec{i} - 5 \sin 8 \, \vec{j})$
    $= 1.8189\ldots \vec{i} + 2.7532\ldots \vec{j}$
    Graph, at Problem 8, showing the tangential component of the acceleration vector.

11. The object is slowing down.
    Reasons: (1) The angle between $\vec{a}(4)$ and $\vec{v}(4)$ is larger than 90°.
    (2) The dot product $\vec{v}(4) \cdot \vec{a}(4) = -19.5645\ldots$, which is negative.
    (3) $\vec{a}_t(4)$ points in the opposite direction of $\vec{v}(4)$.
    The object is slowing down at $|\vec{a}_t(4)| = 3.2998\ldots$
    ≈ 3.3 (ft/sec)/sec.

12. $\vec{a}_n(4) = \vec{a}(4) - \vec{a}_t(4) = 1.9650\ldots \vec{i} - 1.2982\ldots \vec{j}$
    Graph, at Problem 8, showing the normal component of the acceleration vector.

13. $\vec{a}_n(4)$ points toward the concave side of the path. This component pulls the object out of a straight-line path into the curved path.

14. [Learn]

Exploration 71: Hypocycloid Vector Problem

1. The grapher graph agrees with the figure.

2. $\vec{r}(t) = (6 \cos 0.5t - \cos 3t)\vec{i}$
   $\qquad\qquad + (6 \sin 0.5t + \sin 3t)\vec{j}$
   $\vec{v}(t) = (-3 \sin 0.5t + 3 \sin 3t)\vec{i}$
   $\qquad\qquad + (3 \cos 0.5t + 3 \cos 3t)\vec{j}$
   $\vec{a}(t) = (-1.5 \cos 0.5t + 9 \cos 3t)\vec{i}$
   $\qquad\qquad + (-1.5 \sin 0.5t - 9 \sin 3t)\vec{j}$

Calculus Explorations
© 1998 Key Curriculum Press

3. $\vec{v}(2) = (-3 \sin 1 + 3 \sin 6)\vec{i} + (3 \cos 1 + 3 \cos 6)\vec{j}$
   $\approx -3.36\,\vec{i} + 4.50\,\vec{j}$
   $\vec{a}(2) = (-1.5 \cos 1 + 9 \cos 6)\vec{i}$
   $\qquad\qquad + (-1.5 \sin 1 - 9 \sin 6)\vec{j}$
   $\approx 7.83\,\vec{i} + 1.25\,\vec{j}$

4. Graph, showing the position, velocity, and acceleration vectors at $t = 2$.

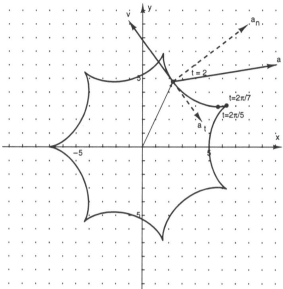

5. The object seems to be slowing down. The angle between $\vec{v}(2)$ and $\vec{a}(2)$ is larger than 90°.

6. $\dfrac{\vec{v}(2) \cdot \vec{a}(2)}{|\vec{v}(2)|} \approx \dfrac{-3.36 \cdot 7.83 + 4.50 \cdot 1.25}{\sqrt{3.36^2 + 4.50^2}} \approx -3.68$

7. The object is slowing down. The scalar projection of $\vec{a}(2)$ onto $\vec{v}(2)$ is negative.

8. $\vec{a}_t(2) = \dfrac{\vec{v}(2) \cdot \vec{a}(2)}{|\vec{v}(2)|^2}\,\vec{v}(2)$
   $\approx \dfrac{-3.36 \cdot 7.83 + 4.50 \cdot 1.25}{3.36^2 + 4.50^2} (-3.36\,\vec{i} + 4.50\,\vec{j})$
   $\approx 2.20\,\vec{i} - 2.95\,\vec{j}$

9. $\vec{a}_n(2) = \vec{a}(2) - \vec{a}_t(2) \approx 5.63\,\vec{i} + 4.20\,\vec{j}$

10. Graph, at Problem 4, showing the tangential and normal components of acceleration.

11. The tangential component is pulling the object in the direction opposite of motion, slowing it down. The normal component is pulling the object to the side, out of a straight path into a curve.

12. At a cusp, $\dfrac{dx}{dt} = \dfrac{dy}{dt} = 0$

13. Technical way:
    Solve $-3 \sin 0.5t + 3 \sin 3t = 3 \cos 0.5t + 3 \cos 3t = 0$.
    Judging by the position of $t = 2$, the cusp near $(6,3)$ will be close to $t = 1$.
    $\sin 0.5t - \sin 3t = 0$, $t \approx 1 \Rightarrow t = 0.8975\ldots$
    $\cos 0.5t + \cos 3t = 0$, $t \approx 1 \Rightarrow t = 0.8975\ldots.$

Check that $\dfrac{dx}{dt}(0.8975\ldots) = \dfrac{dy}{dt}(0.8975\ldots) = 0$ and that $(x(0.8975\ldots), y(0.8975\ldots))$ is near $(6, 3)$. (If the two numerical solutions had been different or the checks had failed, you would have had to solve again using another starting guess for t.)
Intuitive way:
Note the symmetry of the graph. The explanation of the shape (wheels turning within wheels) indicates that the seven cusps are evenly spaced; and the $\vec{j}$-component of $\vec{r}$ is an odd function—in particular $\vec{r}(0)$ is on the x-axis midway between two cusps. Therefore, the first of seven cusps should be at $t = \dfrac{4\pi}{14} = 0.8975\ldots.$

Check that $\dfrac{dx}{dt}(0.8975\ldots) = \dfrac{dy}{dt}(0.8975\ldots) = 0$ and that $(x(0.8975\ldots), y(0.8975\ldots))$ is near $(6, 3)$.

14. Solve $3 \cos 0.5t + 3 \cos 3t = 0$ with an initial guess of about 1.2, with a range of $0.9 \le t \le 2$:
    $t = 1.2566\ldots$ (Exactly $t = 0.4\pi$)

15. Speed $= |\vec{v}(2)| \approx \sqrt{3.36^2 + 4.50^2} \approx 5.62$ ft/sec

16. $dL = \sqrt{x'(t)^2 + y'(t)^2}\; dt = |\vec{v}(t)|\; dt$
    $= [(-3\sin 0.5t + 3\sin 3t)^2$
    $\qquad\qquad + (3\cos 0.5t + 3\cos 3t)^2]^{1/2}\; dt$
    $= 3\sqrt{2\cos 0.5t \cos 3t + \cos^2 3t}\; dt$
    $= 3\sqrt{2 + 2(\cos 0.5t \cos 3t - \sin 0.5t \sin 3t)}\; dt$
    $= 3\sqrt{2 + 2\cos 3.5t}\; dt$
    $= 3\sqrt{2 + 2(2\cos^2 1.75t - 1)}\; dt$
    $= 3\sqrt{4\cos^2 1.75t}\; dt$
    $= 6\,|\cos 1.75t|\; dt$
    One complete cycle is $0 \le t \le 4\pi$, so
    $L = \displaystyle\int_{t=0}^{t=4\pi} dL$
    $= \displaystyle\int_{0}^{4\pi} 6\,|\cos 1.75t|\; dt$
    $= 14 \displaystyle\int_{0}^{2\pi/7} 6 \cos 1.75t\; dt \qquad$ (by symmetry)
    $= \dfrac{84}{1.75} \sin 1.75t \Big|_{0}^{2\pi/7}$
    $= \dfrac{84}{1.75} \sin \dfrac{\pi}{2}$
    $= 48$ ft.
    This is reasonably close to the circumference of a 7-foot-radius circle, $14\pi = 42.9822\ldots$ ft.

17. [Learn]

The Calculus of Variable-Factor Products

Exploration 72: Pumping Work Problem

1. The volume of the slice is $dV = \pi x^2\, dy$,
   so the weight $= 63 \cdot \pi x^2\, dy$.
   The distance to lift the slice is $15 - y$,
   so $dW = 63\pi(15 - y)x^2\, dy$.

2. The radius x of the vat at height y is 6 ft at $y = 0$;
   also, x is 8 ft at $y = 10$, and varies linearly with y.
   Therefore,
   $x = 6 + 0.2y$
   $dW = 63\pi(15 - y)(6 + 0.2y)^2\, dy$
   $\quad = 63\pi(540 - 1.8y^2 - 0.04y^3)\, dy$

3. $W = \displaystyle\int_{y=0}^{y=10} dW$

   $\quad = 63\pi \displaystyle\int_0^{10} 540 - 1.8y^2 - 0.04y^3\, dy$

   $\quad = 63\pi(540y - 0.6y^3 - 0.01y^4)\Big|_0^{10}$

   $\quad = 63\pi(5400 - 600 - 100)$
   $\quad = 296{,}100\pi = 930{,}225.58\ldots$
   $\quad \approx 930$ thousand ft/lb

4. [Learn]

Exploration 73: Mass of a Variable-Density Solid

1. The volume of the slice is
   $dV = \pi x^2\, dy = \pi y\, dy$ (because $y = x^2$).
   $dm = \rho\, dV = 3y^{1/2} \cdot \pi y\, dy = 3\pi y^{3/2}\, dy$
   The mass dm of a slice at height y is
   $dm = 3y^{1/2}\, dV = 3y^{1/2}(2\pi x^2\, dy) = 3y^{1/2}(2\pi y\, dy)$.
   $m = \displaystyle\int_{y=0}^{y=4} dm = 3\pi \displaystyle\int_0^4 y^{3/2}\, dy$

   $\quad = \tfrac{6}{5}\pi y^{5/2}\Big|_0^4$

   $\quad = 38.4\pi = 120.6371\ldots \approx 120.6$ g
   (Numerical integration is okay)

2. The volume of a shell at radius x is
   $dV = 2\pi x(4 - y)\, dx = 2\pi x(4 - x^2)\, dx$.
   The mass dm of the shell is
   $dm = (x + 5)\, dV = (x + 5) \cdot 2\pi x(4 - x^2)\, dx$.
   $\quad = 2\pi(20x + 4x^2 - 5x^3 - x^4)\, dx$.
   $m = \displaystyle\int_{x=0}^{x=2} dm = 2\pi \displaystyle\int_0^2 (20x + 4x^2 - 5x^3 - x^4)\, dx$.

   $\quad = 2\pi(10x^2 + \tfrac{4}{3}x^3 - \tfrac{5}{4}x^4 - \tfrac{1}{5}x^5)\Big|_0^2$

   $\quad = 48\tfrac{8}{15}\pi = 152.4719\ldots \approx 152.5$ g

3. [Learn]

Exploration 74: Moment of Volume, and Centroid

1. $dV = \pi x^2\, dy = \pi y\, dy$

   $V = \pi \displaystyle\int_0^4 y\, dy = \tfrac{\pi}{2}y^2\Big|_0^4 = 8\pi$

   Or just recall $V =$ one-half the volume of the

---

circumscribed cylinder.)

2. Within the disk, the displacement y from the
   xz-plane remains essentially constant.
   So $dM_{xz} = y\, dV$.

3. $dV = \pi y\, dy$
   $dM_{xz} = \pi y^2\, dy$

   $M_{xz} = \displaystyle\int_{y=0}^{y=4} dM_{xz} = \displaystyle\int_0^4 \pi y^2\, dy = \tfrac{\pi}{3}y^3\Big|_0^4 = \tfrac{64}{3}\pi$

4. $M_{xz} = \overline{y} \cdot V \Rightarrow \overline{y} = \tfrac{8}{3} = 2.6666\ldots$

   Graph, showing the centroid. This point is more
   than halfway to the top because there is more
   volume at the top than at the bottom.

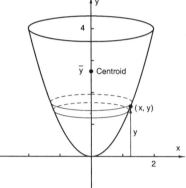

5. [Learn]

Exploration 75: Moment of Mass and Center of Mass

1. The volume of the slice is
   $dV = \pi x^2\, dy = \pi y\, dy$ (because $y = x^2$).
   $dm = \rho\, dV = 3y^{1/2} \cdot \pi y\, dy = 3\pi y^{3/2}\, dy$
   The mass dm of a slice at height y is
   $dm = 3y^{1/2}\, dV = 3y^{1/2}(2\pi x^2\, dy) = 3y^{1/2}(2\pi y\, dy)$.
   $m = \displaystyle\int_{y=0}^{y=4} dm = 3\pi \displaystyle\int_0^4 y^{3/2}\, dy$

   $\quad = \tfrac{6}{5}\pi y^{5/2}\Big|_0^4$

   $\quad = 38.4\pi = 120.6371\ldots \approx 120.6$ g
   (Numerical integration is okay)

2. Within the disk, the displacement y from the
   xz-plane remains essentially constant. So
   $dM_{xz} = y\, dm$.

3. $dm = 3y^{1/2}\, dV = 3\pi y^{3/2}\, dy$
   $dM_{xz} = 3\pi y^{5/2}\, dy$

   $M_{xz} = \displaystyle\int_{y=0}^{y=4} dM_{xz} = \displaystyle\int_0^4 3\pi y^{5/2}\, dy$

   $\quad = \tfrac{6\pi}{7}y^{7/2}\Big|_0^4 = \tfrac{768}{7}\pi$

   $\quad = 109\tfrac{5}{7}\pi = 344.6775\ldots \approx 344.7$ g $\cdot$ cm

4. $M_{xz} = \overline{y} \cdot m$

$\overline{y} = \dfrac{768\pi}{7} \cdot \dfrac{1}{38.4} = \dfrac{20}{7} = 2.8751\ldots \approx 2.88$ cm

Graph showing the center of mass.

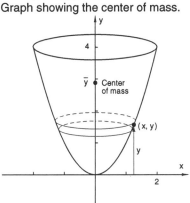

5. The centroid will coincide with the center of mass if the density of the solid is constant.
6. [Learn]

Exploration 76: Second Moment of Mass

1. The volume of the paraboloid is half the volume of the circumscribed cylinder.

$V = \dfrac{1}{2}(\pi \cdot 2^2 \cdot 4) = 8\pi$ cm^3

The density is constant, so the mass is
$m = 2.7 \cdot V = 21.6\pi = 67.8584\ldots \approx 67.9$ g.

2. Within each shell, the distance to the y-axis remains essentially constant. So $dM_{2y} = x^2\, dm$.

3. $dM_{2y} = x^2\, dm = x^2 \cdot 2.7\, dV = 2.7x^2 \cdot 2\pi x(4-y)\, dx$
$= 5.4\pi x^3(4 - x^2)\, dx$

$M_{2y} = \displaystyle\int_{x=0}^{x=2} dM_{2y} = \pi \int_0^2 21.6x^3 - 5.4x^5\, dx$

$= 5.4\pi x^4 - 0.9\pi x^6 \big|_0^2$

$= 28.8\pi$ g $\cdot$ cm^2

4. $M_{2y} = \overline{r}^2 \cdot m \Rightarrow \overline{r}^2 = \dfrac{4}{3}$ cm$^2 \Rightarrow$

$\overline{r} = \dfrac{2\sqrt{3}}{3} = 1.1547\ldots \approx 1.15$ cm

5. The second moment of mass is the mass times the *square* of the distance from the axis, so moving the weight three times as far from the axis will increase the second moment of mass by a factor of $3^2 = 9$.

6. [Learn]

Exploration 77: Force Exerted by a Variable Pressure

1. $p = 62.4(12 - y)$
2. $dF = p\, dA = 62.4(12 - y) \cdot (2x\, dy)$
3. $dF = 62.4\,(12 - y)\, \dfrac{20}{\sqrt{3}}\, y^{1/2}\, dy$

$= 416\sqrt{3}(12y^{1/2} - y^{3/2})\, dy$

$F = \displaystyle\int_{y=0}^{y=12} dF = 416\sqrt{3} \int_0^{12} (12y^{1/2} - y^{3/2})\, dy$

$= 416\sqrt{3}(8y^{3/2} - 0.4y^{5/2})\big|_0^{12}$

$= 416\sqrt{3} \cdot (38.4\sqrt{12}) = 95{,}846.4$ lb

4. $dM_x = y\, dF = 416\sqrt{3}(12y^{3/2} - y^{5/2})\, dy$

$M_x = 416\sqrt{3} \displaystyle\int_0^{12} (12y^{3/2} - y^{5/2})\, dy$

$= 416\sqrt{3}(4.8y^{5/2} - \tfrac{2}{7}y^{7/2})\big|_0^{12}$

$= 416\sqrt{3}\,(4.8 \cdot 12^{5/2} - \tfrac{2}{7} \cdot 12^{7/2})$

$= \dfrac{3{,}450{,}470.4}{7} = 492{,}924\tfrac{24}{70}$ ft $\cdot$ lb

5. $M_x = \overline{y} \cdot F \Rightarrow \overline{y} = \dfrac{36}{7} = 5\tfrac{1}{7}$ ft

6. [Learn]

Exploration 78: Spindletop Oil Well Problem

1. $\dfrac{dB}{dt} = 100{,}000 \cdot 2^{-t/9}$

2. $B = \displaystyle\int_{t=0}^{t=9} dB = 100{,}000 \int_0^9 2^{-t/9}\, dt$

$= 100{,}000 \cdot \dfrac{-9}{\ln 2} \cdot 2^{-t/9}\big|_0^9$

$= \dfrac{-900{,}000}{\ln 2}\,(2^{-1} - 1)$

$= \dfrac{450{,}000}{\ln 2}$

$= 649{,}212.7684\ldots$

≈ 649 thousand barrels

3. Assume that oil is worth \$24 per barrel. Worth is $(649{,}212.7684\ldots)(24)$.
$\approx$ \$15.6 million!

4. $(649{,}212.7684\ldots) \backslash f(42.5, 44{,}000) \approx 627$ tank cars
Length $\approx 627(60) = 37{,}620$ ft
or about 7.1 miles!

5. (See the referenced book.)

6. [Learn]

Exploration 79: Two Geometric Series

1. hours mg
     0   30
     1   27
     2   24.3
     3   21.87
     4   19.683

2. $30 + 30(0.9) + 30(0.9)^2 + 30(0.9)^3 + 30(0.9)^4$
   $= 30(1 + (0.9) + (0.9)^2 + (0.9)^3 + (0.9)^4)$
   $= 30\dfrac{1 - 0.9^5}{1 - 0.9}$
   $= 30 \cdot 4.0951$
   $= 122.853$ mg

3. Answer will depend on the particular grapher used.
   For a TI-83 the commands are
   `sum(seq(30*0.9^N,N,0,4))`
   and the result is 122.853 as in Problem 2.

4. At t = 10 hours,
   $30 + 30(0.9) + \ldots + 30(0.9)^{10} = 205.8568\ldots$
   ≈ 206 mg.
   At t = 20 hours,
   $30 + 30(0.9) + \ldots + 30(0.9)^{20} = 267.1743\ldots$
   ≈ 267 mg.

5. The partial sums converge to 300 mg. The patient will never have more than 300 mg of antibiotic in her body.

6. After 10 years,
   $800 + 800(1.1) + 800(1.1)^2 + \ldots + 800(1.1)^{10}$
   $\approx \$14{,}824.93.$
   After 20 years,
   $800 + 800(1.1) + 800(1.1)^2 + \ldots + 800(1.1)^{20}$
   $\approx \$51{,}202.00.$
   After 30 years,
   $800 + 800(1.1) + 800(1.1)^2 + \ldots + 800(1.1)^{30}$
   $\approx \$145{,}554.74.$

7. The totals appear to be diverging to infinity.

8. [Learn]

Exploration 80: A Power Series for a Familiar Function

1. $P_2(0.6) = 1 + 0.6 + \frac{1}{2}0.6^2 = 1.78$

   $P_3(0.6) = 1 + 0.6 + \frac{1}{2}0.6^2 + \frac{1}{6}0.6^3 = 1.816$

   $P_4(0.6) = 1 + 0.6 + \frac{1}{2}0.6^2 + \frac{1}{6}0.6^3 + \frac{1}{24}0.6^4 = 1.8214$

2. | n | $P_n(0.6)$ |
   |---|---|
   | 5 | 1.822048 |
   | 6 | 1.8221128 |
   | 7 | 1.822118354285... |
   | 8 | 1.822118770857... |
   | 9 | 1.822118798628... |
   | 10 | 1.822118800294... |

   The partial sums seem to be approaching 1.8221188.... (Exactly $e^{0.6}$)

3. Graph, $P_{10}(x) = 1 + x + \frac{1}{2!}x^2 \ldots + \frac{1}{10!}x^{10}$.

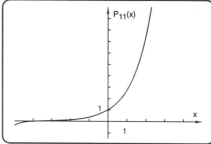

4. $P_{10}(1) = 1 + 1 + \frac{1}{2!} \ldots + \frac{1}{10!} = 2.71828180\ldots$
   $P_{10}(1)$ is very close to e = 2.7182...!

5. Conjecture: $P(x) \approx e^x$ when x is close to zero.
   Evidence: $P_{10}(1)$ is very close to e^1.
   $P_{10}(0.6)$ is very close to $e^{0.6} = 1.8221\ldots$.
   $P_{10}(0) = 1$, which equals e^0.
   A table of $P_{10}(x)$ and e^x shows the values are almost indistinguishable as long as x is reasonably close to zero.
   The graph of $P_{10}(x)$ in Problem 3 resembles the graph of $y = e^x$.
   For most n, $P_n(x)$ is similar to $P_n'(x) = P_{n-1}'(x)$.

6. $e^{10} = 22{,}026.4657\ldots$
   $P_{10}(10) = 12{,}842.3051\ldots,$
   which is *not* close to e^{10}.

7. [Learn]

Exploration 81: Power Series for Other Familiar Functions

1. $f(x) = x - \frac{1}{3!}x^3 + \frac{1}{5!}x^5 - \frac{1}{7!}x^7 + \frac{1}{9!}x^9 - \frac{1}{11!}x^{11} + \ldots$
   Graph, $f_4(x) = x - \frac{1}{3!}x^3 + \frac{1}{5!}x^5 - \frac{1}{7!}x^7 + \frac{1}{9!}x^9$.
   Conjecture: $f(x) = \sin x$

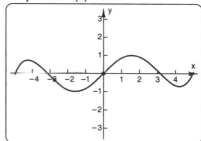

2. $f(0) = 0$
   $f'(x) = 1 - \frac{3}{3!}x^2 + \frac{5}{5!}x^4 - \frac{7}{7!}x^6 + \ldots$
   $\quad = 1 - \frac{1}{2!}x^2 + \frac{1}{4!}x^4 - \frac{1}{6!}x^6 + \ldots$
   $f'(0) = 1$
   $f''(x) = 0 - \frac{2}{2!}x^1 + \frac{4}{4!}x^3 - \frac{6}{6!}x^5 + \ldots$
   $\quad = -x + \frac{1}{3!}x^3 - \frac{1}{5!}x^5 + \ldots$
   $f''(0) = 0$

   $f'''(x) = -1 + \frac{3}{3!}x^2 - \frac{5}{5!}x^4 + \ldots =$
   $$-1 + \frac{1}{2!}x^2 - \frac{1}{4!}x^4 + \ldots$$
   $f'''(0) = -1$
   $f^{(4)}(x) = 0 + \frac{2}{2!}x^1 - \frac{4}{4!}x^3 + \ldots = -\frac{1}{3!}x^3 + \ldots$

$f^{(4)}(0) = 0$
$\sin 0 = 0$
$\sin' 0 = \cos 0 = 1$
$\sin'' 0 = -\sin 0 = 0$
$\sin''' 0 = -\cos 0 = -1$
$\sin^{(4)} 0 = \sin 0 = 0$
sin 0 and the first four derivatives of sin x match those of f(x) at x = 0!
Also, note that both sin x and f(x) are odd functions.

3. $g(x) = 1 - \frac{1}{2!}x^2 + \frac{1}{4!}x^4 - \frac{1}{6!}x^6 + \frac{1}{8!}x^8 - \frac{1}{10!}x^{10} + \ldots$

   Graph, $g_4(x) = 1 - \frac{1}{2!}x^2 + \frac{1}{4!}x^4 - \frac{1}{6!}x^6 + \frac{1}{8!}x^8$.
   Conjecture: $g(x) = \cos x$

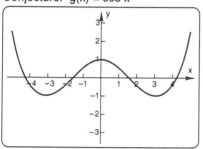

4. $h(x) = 1 + \frac{1}{2!}x^2 + \frac{1}{4!}x^4 + \frac{1}{6!}x^6 + \frac{1}{8!}x^8 + \frac{1}{10!}x^{10} + \ldots$

   Graph, $h_4(x) = 1 + \frac{1}{2!}x^2 + \frac{1}{4!}x^4 + \frac{1}{6!}x^6$.
   Conjecture: $h(x) = \cosh x$

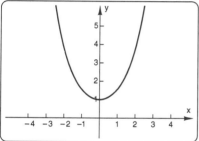

5. [Learn]

Exploration 82: A Power Series for a Definite Integral

1. $V = \int_0^\pi 2\pi x \cdot (x \sin x)\, dx = 2\pi \int_0^\pi x^2 \sin x\, dx$

2. $x^2 \sin x = x^2 (x - \frac{1}{3!}x^3 + \frac{1}{5!}x^5 - \frac{1}{7!}x^7 + \frac{1}{9!}x^9 \ldots)$

   $= x^3 - \frac{1}{3!}x^5 + \frac{1}{5!}x^7 - \frac{1}{7!}x^9 + \frac{1}{9!}x^{11} - \ldots$

3. $V = 2\pi \int_0^\pi (x^3 - \frac{1}{3!}x^5 + \frac{1}{5!}x^7 - \frac{1}{7!}x^9 + \frac{1}{9!}x^{11} - \ldots)\, dx$

   $= 2\pi(\frac{1}{4}x^4 - \frac{1}{6}\cdot\frac{1}{3!}x^6 + \frac{1}{8}\cdot\frac{1}{5!}x^8 - \frac{1}{10}\cdot\frac{1}{7!}x^{10} + \ldots)\big|_0^\pi$

   $= 2\pi(\frac{1}{4}\pi^4 - \frac{1}{6}\cdot\frac{1}{3!}\pi^6 + \frac{1}{8}\cdot\frac{1}{5!}\pi^8 - \frac{1}{10}\cdot\frac{1}{7!}\pi^{10} + \ldots)$

   $= 2\pi(\frac{1}{4}\pi^4 - \frac{1}{6}\cdot\frac{1}{3!}\pi^6 + \frac{1}{8}\cdot\frac{1}{5!}\pi^8 - \frac{1}{10}\cdot\frac{1}{7!}\pi^{10} + \ldots)$

   $V \approx 36.8743\ldots$

4. $2\pi \int_0^\pi x^2 \sin x\, dx$

   $= 2\pi(-x^2 \cos x + 2x \sin x + 2 \cos x)\big|_0^\pi$

   $= 2\pi(\pi^2 - 4) = 36.8798\ldots$

The estimate from Problem 3 is very close to the actual answer, only about 0.015% off.

5. [Learn]

Exploration 83: Introduction to the Ratio Technique

1. $\ln x = (x - 1) - \frac{1}{2}(x - 1)^2 + \frac{1}{3}(x - 1)^3 - \frac{1}{4}(x - 1)^4 + \ldots$

2.

n	t_n	$|t_{n+1}/t_n|$
1	0.6	$0.6 \cdot 1/2 = 0.3$
2	$-0.6^2/2$	$0.6 \cdot 2/3 = 0.4$
3	$0.6^3/3$	$0.6 \cdot 3/4 = 0.45$
4	$-0.6^4/4$	$0.6 \cdot 4/5 = 0.48$
5	$0.6^5/5$	$0.6 \cdot 5/6 = 0.5$
6	$-0.6^6/6$	$0.6 \cdot 6/7 = 0.5142\ldots$
7	$0.6^7/7$	$0.6 \cdot 7/8 = 0.525$
8	$-0.6^8/8$	$0.6 \cdot 8/9 = 0.5333\ldots$
9	$0.6^9/9$	$0.6 \cdot 9/10 = 0.54$

3. See table above.

4. $|t_{n+1}/t_n|$ appears to be approaching 0.6.

5. $|t_{n+1}/t_n| = \frac{n}{n+1} \cdot 0.6$

6. $S_4(1.6) = (1.6 - 1) - \frac{1}{2}(1.6 - 1)^2 + \frac{1}{3}(1.6 - 1)^3$
   $- \frac{1}{4}(1.6 - 1)^4$
   $= 0.6 - 0.18 + 0.072 - 0.0324 = 0.4596$

7. $|t_5| = \frac{1}{5}0.6^5 = 0.015552$

   $|t_6| = \frac{1}{6}0.6^6 = 0.007776$

   $|t_7| = \frac{1}{7}0.6^7 = 0.003999\ldots$

   $|t_8| = \frac{1}{8}0.6^8 = 0.00209952$

   $|t_9| = \frac{1}{9}0.6^9 = 0.0011\ldots$

8. $|t_5| + 0.7|t_5| + 0.7^2|t_5| + 0.7^3|t_5| + 0.7^4|t_5|$
   $= 0.015552(1 + 0.7 + 0.7^2 + 0.7^3 + 0.7^4)$

9. Geometric series converges to
   $\frac{0.015552}{1 - 0.7} = 0.05184$.

10. Each term $|t_n| = \frac{1}{n}0.6^n$ in the tail for $n \geq 5$ is less than $\frac{1}{5}0.6^n$, and hence less than $\frac{1}{5}0.6^5 \cdot 0.7^{n-5}$, which is in turn less than $\frac{1}{5}0.6^5 \cdot 0.7^{n-5} = |t_5| \cdot 0.7^{n-5}$. But then each such term, $|t_n|$ is less than the corresponding term in the geometric series in Problem 8. Therefore, the sum of the absolute values in the tail is less than the sum of the geometric series, 0.05184, so 0.05184 is an upper bound for the sum of the terms in the tail.

11. The terms in the tail of ln 4 have absolute value $\frac{1}{n} \cdot 3^n$; but bounding each term above (as in the preceding problems) by $\frac{1}{5} \cdot 3.1^n$ gives a geometric series which does not converge.

12. $t_n = \frac{1}{n}(x - 1)^n$

    $\left|\frac{t_{n+1}}{t_n}\right| = \left|t_{n+1} \cdot \frac{1}{t_n}\right|$

    $= \left|\frac{(x - 1)^{n+1}}{n + 1} \cdot \frac{n}{(x - 1)^n}\right| = \frac{n}{n+1}|x - 1|$

13. $L = \lim_{n\to\infty} \frac{n}{n+1}|x - 1| = |x - 1| \lim_{n\to\infty} \frac{n}{n+1} = |x - 1|$

14. $L < 1 \Leftrightarrow |x - 1| < 1 \Leftrightarrow -1 < x - 1 < 1$
    $\Leftrightarrow 0 < x < 2$
15. Radius of convergence = 1
16. [Learn]

Exploration 84: Improper Integrals to Test for Convergence

1. R_4 is the remainder starting at t_5. As shown on the graph, if rectangles are drawn with their right-hand sides touching the graph, they will be inscribed in the region under the graph. The left-hand side of the rectangle for t_5 is at n or x = 4. Thus the integral from 4 to infinity is an upper bound for R_4, and hence R_4 is a lower sum for the integral.

2. $\int_4^{\infty} x^{-2}\, dx = -x^{-1}\big|_4^{\infty} = \frac{1}{4}$

3. $t_5 = \frac{1}{25} = 0.04$

   $t_5 + t_6 = \frac{1}{25} + \frac{1}{36} = 0.0677\ldots$

   $t_5 + t_6 + t_7 = \frac{1}{25} + \frac{1}{36} + \frac{1}{49} = 0.0881\ldots$

4. Although the terms are decreasing, they are still positive. Adding positive terms produces increasing partial sums.

5. By definition, if the sequence of partial sums converges, the series converges. Since the sequence of partial sums for R_4 converges, so does the series.

6. By the closure axiom for addition of real numbers, any partial sum is a real number. If the remainder is also a real number, then the entire series converges.

7. The sequence of partial sums is increasing, but now the partial sums are not bounded above but rather bounded below by the value of the integral, so the property in Problem 5 cannot be used. Indeed, saying that a series is larger than some finite number does not prevent the series from being infinite!

8. [Learn]

Exploration 85: Convergence of a Series of Constants

1. For a geometric series, $S_n = t_1 \cdot \dfrac{1 - r^n}{1 - r}$.

   For this series, $S_n = 100 \cdot \dfrac{1 - 0.8^n}{1 - 0.8}$.

   Since the limit of 0.8^n is zero as n approaches infinity, S_n approaches 100(1/0.2) = 500 as n approaches infinity.
   Since the sequence of partial sums converges, the series converges by definition.

2. Graph, showing the terms of the series drawn as circumscribed rectangles for f(x) = 1/x, starting at x = 1.

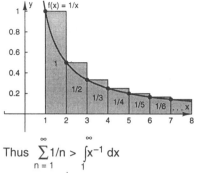

Thus $\displaystyle\sum_{n=1}^{\infty} 1/n > \int_1^{\infty} x^{-1}\, dx$

$= \displaystyle\lim_{b\to\infty} \int_1^b x^{-1}\, dx = \lim_{b\to\infty} (\ln b - \ln 1) = \infty.$

Since the sum is larger than an integral that becomes infinite, the series diverges.

3. $t_5 = \dfrac{256}{40320}$

4. Each term $\dfrac{1}{(2n)!}x^{2n}$ is positive. Adding positive terms increases the sum, regardless of whether the terms are large or small. If the terms being added are getting smaller, then the total sum is still increasing but increasing slowly.

5. $\dfrac{16}{24} + \dfrac{16}{24} \cdot \dfrac{4}{30} + \dfrac{16}{24} \cdot \dfrac{16}{900} + \dfrac{16}{24} \cdot \dfrac{64}{27000} + \cdots$

   $= \dfrac{16}{24} + \dfrac{64}{720} + \dfrac{256}{21600} + \dfrac{1024}{648000} + \cdots$

   Each term of the geometric series has numerator equal to the numerator of a corresponding term in the cosh series, but smaller denominator than that of its corresponding term. Therefore, each term of the geometric series is at least as large as the corresponding term in the cosh series.

6. The geometric series converges to
   $S = \dfrac{16}{24} \cdot \dfrac{1}{1 - 4/30} = \dfrac{10}{13}$.

   The sequence of partial sums for cosh 2 is increasing (see Problem 4), and bounded above by 10/13 (see Problem 5). Thus the series converges.

7. [Learn]

Exploration 86: Introduction to Error Analysis for Series

1. $\cos 0.6 = 1 - \frac{1}{2!}0.6^2 + \frac{1}{4!}0.6^4 - \ldots$
   $= 1 - 0.18 + 0.0054 - \ldots$
   $S_2 = 0.8254$

2. Error $= \cos 0.6 - S_2 = \cos 0.6 - 0.8254$
   $= -0.0000643850903\ldots$

3. $|t_3| = \left| -\frac{1}{6!}0.6^6 \right| = 0.0000648,$

   which is greater than the absolute value of the error shown in Problem 2.

4. The error is also the *remainder* of the series, or the value to which the tail converges.

5. Find t_{n+1} for which $|t_{n+1}| < 0.5 \times 10^{-20}$.

$|t_{n+1}| = \frac{1}{(2n+2)!}0.6^{2n+2}$

$|t_8| = \frac{1}{18!}0.6^{18} = 1.5862\ldots \times 10^{-20}$

$|t_9| = \frac{1}{20!}0.6^{20} = 1.5027\ldots \times 10^{-23}$

Then the error for S_8 (9 terms) is less than $|t_9|$, which is less than 0.5×10^{-20}.
Use 9 terms.

6. $e^{0.6} = 1 + 0.6 + \frac{1}{2!}0.6^2 + \ldots$

$S_2 = 1 + 0.6 + \frac{1}{2!}0.6^2 = 1 + 0.6 + 0.18 = 1.78$

7. Error $= e^{0.6} - S_2 = e^{0.6} - 1.78 = 0.0421\ldots$

8. $t_3 = \frac{1}{3!}0.6^3 = 0.036$

Error is $\frac{0.0421\ldots}{0.036} = 1.1699\ldots$ times the value of t_3

$F = 1.1699\ldots$

9. If $y = e^x$, then $y''' = e^x$ (as do all derivatives of y). Since e^x is an increasing function, the maximum value on [0, 0.6] is $e^{0.6} = 1.8221\ldots$ at the right endpoint.
$F = 1.1699\ldots < 1.8221\ldots = $ maximum y'''.

10. In the Lagrange form, error $= \frac{f^{(n+1)}(c)}{(n+1)!} \cdot x^{n+1}$ for

some point $c \in [0, x]$.
Thus error $< \frac{M}{(n+1)!} \cdot x^{n+1}$,

where M is the maximum value of $f^{(n+1)}$ in [0, x].
In this case, Problems 6–9 have demonstrated that

Error $= F \cdot t_3 = F \cdot \frac{1}{(2+1)!} \cdot x^{2+1} < \frac{e^{0.6}}{(2+1)!} \cdot x^{2+1}$

Thus the work in Problems 6 through 9 has demonstrated the correctness of the Lagrange form of the remainder.

11. [Learn]

Exploration 87: Error Analysis by Improper Integral

1. $S_4 = 1 + \frac{1}{4} + \frac{1}{9} + \frac{1}{16} = 1.42361111\ldots$

2. Upper bound $= \int_5^\infty (x-1)^{-2}\, dx$

$= -(x-1)^{-1}\Big|_5^\infty = \frac{1}{4}$

3. Lower bound $= \int_5^\infty x^{-2}\, dx$

$= -x^{-1}\Big|_5^\infty = \frac{1}{5}$

4. The error is greater than 0.2, so the estimate is not very good. But the error is also less than 0.25, so neither is the estimate very bad.

5. $S_{100} = 1 + \frac{1}{4} + \frac{1}{9} + \ldots + \frac{1}{10000} = 1.63498390\ldots$

Upper bound for error $= \int_{101}^\infty (x-1)^{-2}\, dx$

$= \lim_{x \to \infty} -(x-1)^{-1}\Big|_{101}^b$

$= \frac{1}{100} = 0.01$

Lower bound for error $= \int_{101}^\infty x^{-2}\, dx$

$= \lim_{x \to \infty} -x^{-1}\Big|_{101}^b$

$= \frac{1}{101} = 0.0099009900\ldots$

Thus S_{100} is within 0.01 of the limit, or within ±1 in the second decimal place.
A good estimate would be S_{100} plus half of 0.01.
$S_{100} + 0.005 = 1.634983\ldots + 0.005 = 1.639983\ldots$
≈ 1.640
Note that you are not justified in keeping digits beyond the third decimal place.

6. [Learn]

Calculus Correlation Index

Main Topic	Exploration Number	Main Topic	Exploration Number
Acceleration, from displacement	15	Derivative, parametric function	24
Acceleration, from velocity	15, 19	Derivative, power	14
Acceleration, normal component	70, 71	Derivative, product	20
Acceleration, tangential component	70, 71	Derivative, quotient	21
Acceleration, vector	69, 70	Derivative, review	11
Antiderivative	19, 26	Derivative, sine and cosine	16, 18
Antiderivative, exponential function	38	Differentiability implies continuity	23
Arc length	53	Differential	27
Area in polar coordinates	55	Differential equation	40, 42, 43, 44, 45, 46
Area, surface of revolution	54		
Average velocity	66	Differentiation, implicit	25
Average velocity vector	70	Differentiation, logarithmic	38
Axial variation	73, 75	Discontinuity, concept	7
Base e logs vs. natural logs	39	Displacement from acceleration	65
Base of natural logarithms	39	Displacement, from velocity	19
Catenary	62	Displacement, from velocity	26, 33
Center of mass	75	Divergent series	79
Center of pressure	77	Doubly-curved surface	54
Center of volume	See centroid.	e	39
Centroid	74	Error analysis by improper integral	87
Chain experiment	62	Error analysis for series	86
Chain rule	16, 17	Euler's method	45
Change of base property	37	Even function	32
Common ratio	79	Exponential function	38
Composite function	17	Exponential function, application	40
Compound interest	40, 42	First moment	74, 75
Concavity	47	Force exerted by a variable pressure	77
Continuity, and differentiability	23	Force, moment of	77
Continuity, concept	7	Frustum of a cone	54
Convergence, integral test	84	Function graphs and slope	2
Convergence, interval of	83	Fundamental theorem	30, 31, 35
Convergence, radius of	83	Geometric series	79, 85
Convergence, series of constants	84, 85	Gyration, radius of	76
Convergent series	79	Harmonic series	85
Cusp	7, 71	Heaviside's method	61
Cylindrical shells	52	Hyperbolic function	62
Dam problem	77	Hypocycloid vector problem	71
Definite integral, application	33, 72, 73, 74, 75, 76, 77, 78	Implicit relations	25
		Improper integral	63
Definite integral, by power series	82	Improper integral, error analysis by	87
Definite integral, concept	3	Indefinite integral	19
Definite integral, exact value	31	Inertia, moment of	76
Definite integral, properties	32	Initial condition	19, 42
Definite integral, review	11	Instantaneous rate	1
Density, variable	73	Integral test for convergence	84
Derivative, concept	1	Integral, from graph	48
Derivative, exact value	12	Integration techniques	64
Derivative, exponential function	38	Integration, by reduction formula	57
Derivative, from definition	12	Integration, partial fractions	61
Derivative, from graph	48	Integration, parts	56
Derivative, inverse trig	22	Integration, powers of trig functions	58
Derivative, numerical	13	Integration, special trig forms	59
Derivative, of a definite integral	35		

Main Topic	Exploration Number	Main Topic	Exploration Number
Integration, trig substitution	60	Ratio test	*See ratio technique.*
Interval of convergence	83		
Inverse trig functions	22	Rational algebraic function, integration	61
l'Hospital's rule	41		
Length of a plane curve	53	Reduction formula	57
Limit, concept	5	Regression analysis	15
Limit, definition	6, 10	Related rates	67
Limit, exact value	41	Remainder of a series	84
Limit, involving infinity	9	Removable discontinuity	5
Limit, properties	7	Riemann sum	28, 30
Limit, review	11	Sample point	28
Linearization of a function	27	Sandwich theorem	*See squeeze theorem.*
Local linearity	27		
Logarithm, natural	36	Scalar projection	71
Logarithm, properties	37	Second derivative	47
Logarithmic differentiation	38	Second moment of mass	76
Maclaurin series	85	Separation of variables	42
Mass, center of	75	Series, convergent	79
Mass, moment of	75	Series, divergent	79
Mass, second moment	76	Series, error analysis	86
Mass, variable density	73	Series, geometric	79, 85
Mathematical induction	7	Series, harmonic	85
Maxima and minima	47, 49	Series, Maclaurin	85
Maximal cylinder problem	49	Series, of constants	84, 85
Mean value theorem	29	Series, *p*-series	84
Memory retention problem	43	Series, power	80, 81
Minimal path	68	Series, remainder of	84
Moment, of force	77	Series, Taylor	83
Moment, of inertia	76	Sigma notation	81
Moment, of mass	75	Simpson's rule	34
Moment, of volume	74	Singly-curved surface	54
Natural logarithm	36	Slope field	44, 45, 46
Normal component of acceleration	70, 71	Special trig integrals	59
Odd function	32	Speed	70
p-series	84	Spindletop oil well problem	78
Parametric function	24, 41	Squeeze theorem	18
Partial fractions	61	Sum, partial	80
Partial sum	80	Sum, Riemann	28, 30
Parts, integration by	56	Surface of revolution	41
Path, minimal	68	Tangential component of acceleration	70, 71
Plane curve	53		
Plane slices	50, 51	Target function	81
Point of inflection	47	Taylor series	83
Polar coordinates	55	Trapezoidal rule	4
Position vector	70	Trigonometric substitution	60
Power rule	14, 20	Uniqueness theorem for derivatives	36
Power series	80, 81	Variable density mass	73
Power series, for definite integral	82	Variable pressure	77
Powers of trig functions, integration	58	Vector projection	71
Predator-prey problem	46	Vector, position	70
Pressure, center of	77	Vectors	69, 70
Pressure, variable	77	Velocity, average	66
Pumping work problem	72	Velocity, from displacement	15
Quotient rule	21	Velocity, vector	69, 70
Radial variation	73	Volume, by cylindrical shells	52
Radius of convergence	83	Volume, by plane disk	50
Radius of gyration	76	Volume, by plane washer	51
Rates, related	67	Volume, moment of	74
Ratio technique	83	Work, pumping problem	72